SAFETY COMMUNICATIONS

BY
MITCH SCHNEIDER

THOMSON
DELMAR LEARNING

Australia • Canada • Mexico • Singapore • Spain • United Kingdom • United States

THOMSON
DELMAR LEARNING™

Automotive Service Management

Safety Communications

Mitch Schneider

Business Unit VP
Alar Elken

Executive Director, Professional Business Unit
Greg Clayton

Product Development Manager
Timothy Waters

Editorial Assistant
Kristen Shenfield

Development
Kristen Shenfield

Production Director
Mary Ellen Black

Production Manager
Larry Main

Production Editor
Elizabeth Hough

Channel Manager
Beth A. Lutz

Marketing Coordinator
Brian McGrath

Cover Design
Julie Lynn Moscheo

Printed in Canada
1 2 3 4 5 6 7 8 9 XXX 07 06 05 04 03

For more information contact Delmar Learning
Executive Woods
5 Maxwell Drive, PO Box 8007
Clifton Park, NY 12065-8007
Or find us on the World Wide Web at
http://www.delmar.com or **http://www.autoed.com**

Library of Congress Cataloging-in-Publication Data

Schneider, Mitch.
Automotive service management. Safety communications / Mitch
Schneider.-- 1st ed.
p. cm.
Includes bibliographical references.
ISBN 1-4018-2662-8
1. Service stations--Management. 2. Safety education. 3. Industrial accidents--Prevention. I. Title: Safety communications. II. Title.
TL153.S36235 2003
629.28'6'068--dc21
2003004313

NOTICE TO THE READER

Publisher does not warrant or guarantee any of the products described herein or perform any independent analysis in connection with any of the product information contained herein. Publisher does not assume, and expressly disclaims, any obligation to obtain and include information other than that provided to it by the manufacturer.

The reader is expressly warned to consider and adopt all safety precautions that might be indicated by the activities herein and to avoid all potential hazards. By following the instructions contained herein, the reader willingly assumes all risks in connection with such instructions.

The publisher makes no representation or warranties of any kind, including but not limited to, the warranties of fitness for particular purpose or merchantability, nor are any such representations implied with respect to the material set forth herein, and the publisher takes no responsibility with respect to such material. The publisher shall not be liable for any special, consequential, or exemplary damages resulting, in whole or part, from the readers' use of, or reliance upon, this material.

CONTENTS

PREFACE

This shop operations guide is designed to provide a framework that helps you make consistent, high-quality, and productive service a part of your normal, everyday shop operations.

Is it everything you ever wanted to know about automotive shop management but were afraid to ask?

I'm not sure. In reality, it is a lot of what I have learned about running a successful automotive service business, both on my own and sitting at the feet of those who started this journey before I did. At the very least, it's everything you can fit into eight volumes.

The purpose of this guide is to help ensure that *great performance* is achieved every time you or one of your employees approaches a regular or potential customer to deliver automotive service. It is the only way I know to create trust and insure customer loyalty. And, great performance coupled with increased customer loyalty, trust, and operational excellence will almost always result in increased profits.

High-quality service is defined by the customer's experience while in your care and custody. Operational excellence ensures that the customer's cars will be *ready when promised* and *fixed right the first time.* Fixing the car right the first time is as much about understanding what your customer is trying to tell you when they are communicating their frustrations with the vehicle as it is about proper diagnosis and professional quality work. *Total Quality Service* is only possible through high-quality communication and knowing where you are in the process of completing a job at all times. Because the workforce is diverse, we chose to alternate between "he" and "she" from chapter to chapter to ensure gender is equally represented.

Selling service is perhaps the most difficult of all sales to make. Buying a product—something you can hold in your hand like an oil filter or a spark plug—constitutes the purchase of something you can touch or feel. Purchasing a service is something else entirely. When a vehicle owner buys a service, that person is really purchasing a promise—a promise that will be fulfilled in the future. That requires trust and a great deal of faith.

Your ability to perform is based upon a lot of things, not the least of which is technical competency. But you can't demonstrate that technical competency unless (or until) you are given the opportunity. The vehicle owner can't see technical competency. She can't touch it or feel it and you can't hang it out in front of the building like a sign.

The customer's perception of whether or not you are worthy of her trust, and whether or not you will be able to solve her automotive service problem, will be based in many cases upon any one of a number of subtle, almost subliminal, messages. These messages are broadcast and received before any one of your technicians ever picks up a screwdriver, wrench, ratchet, or socket.

Service starts when a customer sees and responds to an advertisement (an invitation to experience the quality of your service), hears about your business from a friend or relative, calls, or walks in. Service includes every subsequent customer contact until the job is completed—ready when promised. Moments of truth—moments when a customer has the opportunity to come in contact with any member of your organization and form an opinion about you or your company—occur before, during, and after the job has been completed. They will ultimately define your personal success and the success of your company.

Consistent execution of company policies and procedures is the catalyst for world-class service. It is perhaps more important to be consistent at what you do than to be good at what you do. Do a great job one time, and something less than great the next, and you aren't likely to see that customer again. The more people you have working for you, the greater the challenge is to achieve consistency. Without a comprehensive and complete set of policies and procedures, consistency is nothing more than an elusive dream.

All of this consistent high-quality service is only sustainable if your shop is operating at productivity levels that allow you to make a profit. Productivity is a by-product of your technicians' ability to execute policies and procedures flawlessly. It will take more than a visit to the tool truck or a one-day management seminar for your technicians to bill 70 percent or more of their available hours.

ABOUT THE SERIES

All eight volumes of this guide—*Total Customer Relationship Management, From Intent to Implementation, Managing Dollars With Sense, Operational Excellence, Building a Team, The High Performance Shop, Safety Communications, and Operations Management*—are designed to work together in large shops and small shops. That means that the building blocks of consistency—the policies and procedures you will find within these pages—are defined in their simplest terms because no matter how big or how small a shop is, the same kinds of tasks must be performed. The policies and procedures we define here can and will be carried out by the service writer, service manager, technician, office manager, and lot person. In many cases, the person who fills the role of service writer also serves as the service manager and perhaps even the technician. The key to success is not who does the job. The key to success is insuring that each critical task is completed accurately and consistently—the same way every time.

While this set of guides will take you through a series of step-by-step procedures, it is important to remember that delivering high-quality service is a process. To achieve success, all of the steps in the process must be performed, either consciously or unconsciously. If you are having a problem in one particular area of your business, start with the guide that addresses that need first. Next, read through the rest of the guides. See if what you find in these pages makes sense to you. *Then, read it again.* Look at your business and see what, if anything, applies to you. *Then, read it again.* Consider which of your current procedures need to remain in place. If some need to be changed, determine if some of the policies or procedures you found in these guides can be incorporated into the daily operation of your business. *Then, read it again.* Put yourself in the place of your customer. Involve everyone in your business in this exercise. You will be far more successful with their help and cooperation than you could ever be without it. Help them understand that this is how things need to be done from *this moment forward* and why. Walk through the new procedures in your shop. Pull the trigger, start the process, read the text, and begin your journey to a better quality of service, a stronger, healthier, more profitable business, and a more satisfying, less frustrating life.

ACKNOWLEDGMENTS

Nothing of consequence occurs in a vacuum. It takes a lifetime of collaboration, cooperation, and support. The Automotive Service Management series is no exception. If it takes a village to raise a child, it takes a city to conceive, write, edit, and publish one book—let alone eight. Consequently, there are some special people I would like to thank for making the journey with me.

I'd like to thank my Delmar family: Tim Waters, Kristen Shenfield, Betsy Hough, and everyone in marketing for staying with me from conception to completion—as well as Lynanne Fowle and Megan Iverson at TIPS Technical Publishing for their meticulous attention to every detail. It is a joy to work with so many people who care.

I'd like to thank John Wick and Willi Alexander for planting the seeds, and Dick Vinet, the Duffy-Vinet Institute, Motor Service Magazine, and Champion Spark Plugs, for presenting the first shop management seminar I ever attended. It gave me confidence, but more than that—it gave me hope. And, Arthur Epstein (of blessed memory) and Ray Walker for their guidance, friendship, and patience and for recognizing in me something I could not or would not see myself.

Most of all I'd like to thank my wife, Lesley, for tolerating a lifetime of ruined dinners, working too late at the shop, and hiding out in the "Cave," writing and thinking and just generally being far away. She's not only been the biggest part of everything I've ever accomplished, believing even when I didn't—she's been the best part of everything I've ever accomplished. I'd like to thank my kids, Ryan and Dana, for sharing me with my work even when I'm sure they didn't understand. They are a blessing and an inspiration.

I'd like to thank my brothers: David for working with me for the better part of thirty-six years, and Todd for not. Both have taught me a great deal. David has taught me about patience and understanding and looking beyond my frustration to try to understand the lesson he is trying to teach me about myself. And, Todd has taught me a great deal about positive attitude, professionalism and personal courage, in addition to never saying no to anything I've ever asked of him.

Finally, I would like to acknowledge my parents—Sylvia and Jerry Schneider. They are far more than parents. They are my closest friends, my favorite teachers, and a living example of what excellence, character, commitment, dedication, sacrifice, and giving everything you have to give is all about. It's hard to relax and take it easy when your eighty-year-old parents won't, and it's just as hard to give less than your best when your eighty-year-old parents won't!

INTRODUCTION

I have been a working automotive service professional for almost thirty-seven years and counting—the better part of my adult life. However, I've been around the industry my entire life. In fact, I was literally born into it: the fourth generation in my family to be so involved.

I climbed the cables on an ancient tow truck in Brooklyn, New York, fantasizing that it was the rigging of an imaginary pirate ship, and stood atop the "mast" while still an elementary school student. I broke tires with a hammer and mounted and dismounted them with two tire irons before I entered junior high. And I walked through bilious clouds of brake dust filled with asbestos fibers originating from brake shoes that were being arced in the shop or brake drums and backing plates that were being blown clean with compressed air well into my twenties, long before anyone realized just how dangerous that stuff really was.

I pumped a few million gallons of leaded gasoline—cleaned parts and washed my hands in it. In fact, I did just about everything but drink it before we finally discovered just how dangerous lead in anything can be.

I hammered on wheel pullers until my arms ached and I used my hips and upper torso as a whip, throwing every ounce I could against the slide hammer, hoping that difficult bearing would finally come loose, never once thinking about safety glasses or a back brace. I used air hammers, ratchets, and air chisels without a thought about ear protection. And I found myself handling carcinogenic chemicals more times than I care to think about right now. The fact of the matter is this: many of us conducted ourselves with reckless abandon, never once giving "hazardous materials" or "safety communication" a second thought.

It was a different world, a different era. Perhaps, as some would suggest, it was an age of innocence. But looking back, I believe it was more appropriately a world of ignorance!

We can whine and complain about government's intrusions into our culture or our industry, or about over-regulation. But realistically, recognizing what we do and the environment in which we do it, I'm sorry there wasn't a little more awareness, intrusion, and over-regulation about thirty-seven years ago when I first starting "wrenching" professionally, and perhaps even before that when I was exposed to any one of a number of different potentially hazardous and equally dangerous situations when I was still just a child.

There is, however, no excuse for such ignorance today. We are more aware of the dangers and hazards that lurk within our walls, both in the dark corners of the shop and inside the office or stockroom, than we have ever been before. And with awareness comes a certain level of responsibility *for* and *to* anyone who enters our world. That responsibility isn't as much about keeping the shop clean, dry, and relatively safe, as it is about education, best practices, responsible policies and procedures, training, and awareness.

Introduction

As long as our environment is filled with dangers and hazards of all kinds, we have a responsibility to communicate the nature of these dangers and hazards to the individuals who spend their lives in our environment working with us and for our customers. We have an equally pressing responsibility to educate and train them in all the different ways there are to avoid these dangers, including safely handling hazardous materials in a careful and responsible way.

We are responsible to ourselves as well, and to our families, to discharge our legal, ethical, and moral responsibilities, especially with regard to communicating the true nature of these hazards and the dangers in the workplace—educating our employees and documenting what materials were handled, how, by whom, and when, as well as what happened to them when they left our care and custody.

That is our minimum responsibility in a world in which you and I are legally liable for the disposition and documentation of those materials and what happened to them from cradle to crypt!

In this volume of the *Automotive Service Management* series, we are going to address these issues as carefully as we can, recognizing that while federal regulations remain consistent for all of us, state and local regulations may vary wildly from city to city and state to state.

If successful, we will create a reference to help you steer your way through the minefield of legislative and administrative requirements and legal and moral obligations, and start you on your way toward a benchmark of best practices when it comes to safety communications and hazardous materials handling, management, and disposal.

CHAPTER 1

Alphabet Soup

INTRODUCTION

This chapter introduces you to:

- The four stages of knowledge

It also gives you an overview of the Occupational Safety and Health Administration requirements with regard to the employee's:

- Right to know
- Right to training
- Right to information
- Right to participate and demand action
- Right of access

In addition, the chapter introduces the following topics for future review:

- An employer's responsibilities
- The concept of hazardous materials communication
- Documentation, record keeping, and training
- Personal protective equipment
- Material handling and storage

AN OVERVIEW

How does anyone make sense out of the alphabet soup of acronyms that populate every government website and are indicative of every government agency? How are you supposed to know what OSHA actually stands for, or what a SIC is, or what NAIC is supposed to do? I found over a thousand government acronyms on one webpage alone: a page whose only purpose was to expand the acronym and nothing more! And once you have found the acronym you are looking for—and if you are fortunate enough to translate it—how do you go about making any sense out of what it is supposed to represent or what your responsibilities really are? How do you make it interesting enough to read and compelling enough to follow through with? Frankly, I'm not sure I know. What I do know, however, is that in many cases the agencies behind the acronyms are regulated by volumes of legislation. I also know that as citizens, employees, and employers, we share a responsibility—a legal responsibility—to comply with the rules and regulations mandated by these federal, state, and local agencies. And I can state with the certainty that comes from personal experience that in most cases this is not a particularly easy thing to do—certainly not if you include it with all the other things we all too often feel we have been compelled or coerced to do.

In the end, working within the guidelines mandated by these agencies and their regulations will seem at some times more worthwhile than at others. For the balance of this book, we will be looking at a number of government agencies and regulations, but we will focus on two in particular: OSHA, the Occupational Safety and Health Administration, and EPA, the Environmental Protection Agency. We will zero in on a number of their regulations and their impact on the workplace, on us, on our industry, on our customers, and on the environment. We are going to look at our legal responsibilities with regard to these agencies as conscientious employers, as reliable employees, and—perhaps most important of all—as responsible citizens. But first, we are going to group these responsibilities into a number of distinct categories.

The first category will focus on the shop owner's responsibilities to employees, customers, and to the community—especially with regard to hazardous materials. The second will focus on responsibly communicating the nature of the materials themselves, the hazards associated with them, as well as how to avoid contamination, and how to contain them responsibly if or when there is a problem. The third and perhaps most important category is associated with the education and training necessary to ensure that everyone likely to come in contact with a hazard of any kind is capable of handling that hazard safely and without incident. The final two areas we will be exploring are the actual handling and storage of hazards or hazardous materials and the use of personal protective equipment.

Our purpose here is to provide you with a brief overview of each of these areas, with a special consideration for how they are all interrelated and interconnected—because, as you will see, there is more than a little overlap. The goal is simple and, interestingly enough, very much the same as OSHA's stated objective, and that is to see "every worker go home whole and healthy every day."

THE FOUR STAGES OF KNOWLEDGE

In *Operations Management*, the sixth book in this series and the one immediately preceding this one, I included the following material on the Four Stages of Knowledge as an example of just how important it is to know and understand where you are in your quest for success. A good deal of that success will be the result of a process of self-actualization—recognizing where you are, where you need to be, what it will take to get there, and the ability to make it all happen. Recognizing what you know and what you don't know will be a big part of that process. There is no one area of your business where this principle is more critical than when it comes to dealing with issues of safety or safety communication.

Here, then, are the Four Stages of Knowledge.

Stage 1

The first stage is simple enough. The individual—or the organization—*does not know what it does not know*. There is no awareness. And consequently, because there is no awareness, this is perhaps the most dangerous stage of all.

Because what is *known* or *not known* is *unknown*, the individual or the organization may attempt to perform in areas in which it is unqualified or ill prepared to perform. Successful task management is impossible if the person responsible for performing the task does not know what he or she does not know. Success can be costly because it may require the application of resources—including time—which the company or the individual may not be able to allocate effectively. Or, in the case of meeting or exceeding the expectations of a governmental regulatory agency such as the Occupational Safety and Health Administration, the result can be devastating in terms of costly fines, citations, litigation, and/or even imprisonment.

Stage 2

The second stage of knowledge—the stage that begins with the realization that *you know what you don't know*—is the beginning of wisdom.

It begins with the realization that you don't know everything, which then allows you to qualify and explore exactly what it is that you need to know. It is the beginning of wisdom because, once you have identified what you need to know, you can take advantage of the available resources to search for the information and the answers necessary to insure your success.

More than that, it is the same kind of self-actualization Abraham Maslow spoke about as the pinnacle of awareness—the highest rung on his ladder of human development—discussed in the fifth book in this series, *Operational Excellence*.

Stage 3

The third stage of knowledge isn't quite as straightforward or easy to understand. It begins with the realization that *you don't know what you know*: that through your research, your education and training—your pursuit of knowledge and experience—you may have learned things that you are unaware that you know.

This is generally the most uncomfortable stage of knowledge because it can result in competent performance or the successful completion of a task without a clear understanding of how or what was done to succeed. It recognizes that not all knowledge is conscious knowledge, and that many times the key to success lies in trusting your instinct or intuition when it tells you it is OK to try.

In many ways, this is also a very dangerous stage of knowledge, because the feeling that you get when instinct or intuition drives you to try something new and different can easily be confused with unearned or undeserved confidence arising from not knowing what you don't know. In fact, the only way I know to tell them apart is to recognize that one is only possible through a constant and total commitment to continual and life-long learning and exploration, and the other is the result of complacency, arrogance, and ignorance.

Stage 4

The fourth and final stage of knowledge occurs when *you know what you know*. It is accompanied by confidence, success, high performance, high self-esteem, and a positive self-image. It is a plateau that you reach after acquiring the knowledge and information necessary for success: a plateau that allows you to prepare for the challenges that accompany the pursuit of new disciplines where you will not know what you don't know yet again.

SHOP RESPONSIBILITIES

There are countless responsibilities facing each of us as business owners and managers today. They range from equal opportunity under the law—especially with regard to discrimination on the basis of race, religion, disability, age, gender, sexual preference, or national origin—to finding skilled workers, or to wellness in the workplace. Each is a subject worthy of at least one book, if not a whole library. And the government agents responsible for monitoring and maintaining compliance in each of these many areas and for their own specific agencies are likely to feel that theirs is—or at least should be—your only agenda. The fact that there is an alphabet soup of agents and agencies doesn't make that any easier; and that's why compliance can be so difficult, confusing, and therefore exhausting.

For instance, the United States Department of Labor, Occupational Safety and Health Administration (USDOL, OSHA) has determined that as an employer you must provide your employees a "workplace free from serious recognized hazards and comply with standards, rules, and regulations issued under the OSHA Act." That means that as an employer it is your responsibility to both know and understand what those rules and regulations are as well as how they will apply to the workplace. It implies a responsibility to ensure workplace conformance to OSHA standards, and that in turn implies that you know and understand what those standards are, and that you will thoroughly examine your place of business to ensure that you are adhering to those standards.

OSHA Standards: What Are They?

OSHA Standards are rules that are designed to protect workers against all the many on-the-job hazards. Standards can relate to the requirement employers face with regard to monitoring hazards in the workplace and/or record keeping. Or they can focus on exposure to dangerous or toxic chemicals. They can mandate the use of personal protective equipment or the need to implement detailed practices in specific situations. The most critical thing you must remember when it comes to OSHA Standards is the fact that as an employer you can be cited and fined if you fail to comply!

OSHA makes it *your* responsibility to ensure that if your employees must use tools and equipment in the performance of their responsibilities on the job, those tools and that equipment must be safe and properly maintained. It also demands that you identify any hazards that might exist in your environment by the use of color codes, posters, labels, and/or signs and mandates education and training with regard to hazards and hazardous materials.

You are required by law to post "at a prominent location within the workplace" the federal OSHA poster or your state's equivalent of that poster informing employees of their rights and responsibilities.

There are very specific regulations demanding documentation of work-related injuries and/or illnesses in companies with ten or more employees. And while your company might be exempted under the federal requirement because you have fewer than ten employees, state rules and regulations may still demand that you keep such a log. If you have a record of illnesses and injuries, remember also that state or federal law is likely to make it your responsibility to share that information with current and/or past employees and their representatives. Federal OSHA regulations also make it the employer's duty to provide access to medical and exposure records for employees and/or their representatives.

In many cases, state or federal rules and regulations will require that someone in management is designated as the company's "compliance officer" and that the person so designated is available to accompany a state or federal OSHA representative in the case of an inspection. And finally, regulations may require that citations are posted publicly at or near the site at which the violation occurred, at least until such time as the violation is corrected, or for up to three days afterward; that violations are addressed immediately and compliance ensured by the deadline date; and that no employee is discriminated against for exercising his rights and responsibilities under the Occupational Safety and Health Administration Act.

OSHA's requirement to create and maintain a "workplace free from serious recognized hazards," one with "standards, rules, and regulations" as issued under the OSHA Act for a safe and hazard free workplace, has generally been interpreted to include "violence in the workplace." And whether you choose to agree or disagree, violence in the workplace has generally been interpreted to include much more than physical violence. Violence in the workplace can include verbal abuse and harassment in addition to physical abuse.

EMPLOYEE RIGHTS

The federal government has dictated that contrary to the way some employers might feel, employees have certain rights. These rights are outlined in the Occupational Safety and Health Act of 1970, and many of these rights are articulated not only in the legislation, but on the OSHA website (*www.osha.gov*) as well.

In business and particularly with regard to employment law, one person's right is another's responsibility.

In this case, we are talking about the employee's right to training. The law indicates that it is not only the *right* of the employee to access training: training is a requirement dictated by OSHA standards. If the employee has the right to access training, training that is required by law, then someone has a *responsibility* to provide that training, and you don't have to be a mind reader to recognize just who that someone might be.

The Right to Know

There was a time, not so long ago, when employers were *not* required by law to inform their employees regarding the dangers, hazards, or possible safety concerns that might be present in the workplace. As a result, employees were regularly put at risk without warning.

Through the establishment of the Occupational Safety and Health Administration and enactment of the Hazard Communication Standard (HCS) 29CFR 1910.1200 and its subsequent revisions, the federal government mandated that employers take specific steps to inform employees about the hazards associated with their work and especially with regard to the materials used in the performance of their job responsibilities.

As you can see in Figure 1-1, the language of the OSHA Right to Know statement that must be posted in a conspicuous location in the workplace is clear:

Figure 1-1 Right to Know poster.

While there are a number of arguments that could be made against government intrusion into the realm of private enterprise, it would be hard to present a legal, moral, or ethical argument against the workers' right to know, understand, recognize, or respond when unnecessarily placed at risk.

The Right to Training

This Right to Know is interpreted in a number of different ways. The statute goes on to clarify the specifics of these rights and requirements when it suggests that the worker has the right to get training from his or her employer in the following areas:

> ... the employee has the right to acquire training from an employer with regard to the chemicals to which they are exposed in the course of performing work-related tasks and information on how to protect themselves from harm.

That right is accompanied by the following responsibility:

> Employers must establish a comprehensive, written hazard communication program.

That means that every employer must see that chemical containers are properly labeled, that Material Safety Data Sheets with detailed hazard information are made available to all employees, and that training is available to ensure that all employees are informed and aware of the risk which accompanies exposure to these chemicals. This should include the long and short term health effects of working with the chemicals as well as the steps that the employer has taken to ensure that everyone in the workplace both knows and understands what he must do to protect himself and everyone else from these hazards.

The training program should provide the employee with a list of all the hazardous chemicals in the workplace and "how the employer will inform employees of the hazards of non-routine tasks (for example, the cleaning of reactor vessels), and hazards associated with chemicals in unlabeled pipes and how the employer will inform other employers at a multi-employer worksite of the hazards to which their employees may be exposed."

It also provides for the right of the employee to get training provided by the employer on a variety of other health and safety hazards.

The Right to Information

Every employee right is accompanied by an employer responsibility. The employee also has a right to know, and that right to know translates to an employer's responsibility to provide information. What does that really mean? It means a lot.

It means that any employee has the right to request information from his employer and that the employer has a legal responsibility to provide that information, especially as it relates to hazards in the workplace, chemicals used or stored in the workplace, and tests the employer has done to monitor and measure levels of toxic chemicals, noise, radiation, and even violence.

The employee has a right to request copies of current OSHA standards, rules, regulations, and requirements. And the employee has a right to access and review what OSHA calls the "Log and Summary of Occupational Injuries and Illnesses (OSHA 30)" at "a reasonable time and in a reasonable manner," or to arrange for an authorized representative to access that information on his behalf. And once again, this translates into a legislative responsibility for garage owners and managers like you and me to have this information available.

The employee should have the right to access and review all relevant exposure and medical records pertaining to hazards in the workplace, and employers are required to communicate what records are being documented, where they are, and how they can be accessed immediately upon employment and then annually thereafter. We'll address this particular right and its accompanying responsibilities a little later on when we talk about documentation, records, and training.

The Right to Participate and Demand Action

As an employee, you have the right to request that your employer will conscientiously address the correction of any workplace hazards or violations, as well as the right to file a confidential complaint if that does not happen in a timely manner. In fact, as an employee, you or one of your designated representatives actually has the right to participate in an OSHA inspection of your place of business.

You also have the right to answer any questions asked of you by an OSHA representative without fear of reprisal or retaliation.

The Right of Access

Employers must make the results of any and all OSHA inspections available to their employees. This is not optional. All employees have the right to review this information without threat of intimidation, reprisal, or retaliation.

If in fact there is any indication that an employee is being "punished" for exercising any of these rights or for refusing to work in a dangerous, hazardous, or in any other way unsafe environment, there is a provision to file a discrimination complaint under Section 11(c) of the OSH Act within thirty days of the occurrence.

EMPLOYER RESPONSIBILITIES

As we have just seen, there are countless responsibilities associated with providing a safe workplace and a hazard-free environment, so many that the burden can seem repressive at times, depending upon the way you have framed them mentally.

These rules and regulations will prove both burdensome and bothersome if you consider the issues trivial and unrelated to your responsibilities as an owner or a manager. If, however, you believe that a big part of your responsibility as an owner or a manager is to provide a safe, hazard-free work environment, you will take both the rules and the responsibilities seriously.

I am going to highlight just of few of these responsibilities for you now. The complete list can be found on the OSHA website at the following web address:

www.osha.gov/as/opa/worker/employer-responsibility.html

But you can rest assured that these topics will continue to be covered throughout the balance of this text in much greater detail as we look at all the responsibilities with regard to a safe and hazard-free shop environment. The owner's responsibilities are to:

- Provide a workplace free from serious recognized hazards and comply with standards, rules, and regulations issued under the OSHA Act
- Examine workplace conditions to make sure they conform to applicable OSHA standards
- Make sure employees have and use safe tools and equipment and properly maintain this equipment
- Use color codes, posters, labels, or signs to warn employees of potential hazards

- Establish or update operating procedures and communicate them so that employees follow safety and health requirements
- Provide medical examinations and training when required by OSHA Standards
- Post, at a prominent location within the workplace, the OSHA poster (or the state plan equivalent) informing employees of their rights and responsibilities
- Report to the nearest OSHA office within eight hours any fatal accident or one that results in the hospitalization of three or more employees
- Keep records of work-related injuries and illnesses (Note: Employers with ten or fewer employees, and employers in certain low-hazard industries, are exempt from this requirement.)
- Provide employees, former employees, and their representatives access to the Log of Work-Related Injuries and Illnesses (OSHA Form 300)
- For 2002 only, post a copy of the totals from the last page of the OSHA 200 Log during the entire month of February 2002
- Provide access to employee medical records and exposure records to employees or their authorized representatives
- Provide to the OSHA compliance officer the names of authorized employee representatives who may be asked to accompany the compliance officer during an inspection
- Post OSHA citations at or near the work area involved. Each citation must remain posted until the violation has been corrected, or for three working days, whichever is longer
- Post abatement verification documents or tags
- Not discriminate against employees who exercise their rights under the Act
- Correct violations by the deadline set in the OSHA citation and submit required abatement verification documentation

HAZARDOUS MATERIALS COMMUNICATION

As you can already begin to see, there are a number of established government standards that identify the rights and define the relationships that exist between government and employer, government and employee, employer and employee, and government and the general public. In many cases, these standards have been legislated by both state and federal government. In fact, twenty-three states currently cover private sector employees and their relationship with their employers, along with state and local government employees, through their own state-run OSHA programs. Central to safety communication is the mandate to ensure that the nature and intent of these standards must be communicated clearly and completely to everyone involved.

Nowhere is this more apparent than in the communication and documentation associated with health and safety issues in the workplace. In other words, if you are in the automotive service business and you have employees, you are required by federal law to communicate these standards, requirements, and responsibilities to everyone in your company. And this is in fact one of the most heavily legislated and demanding responsibilities you will face as a business owner or manager.

com·mu·ni·ca·tion (kə-myoo-ni-kā-shən) *noun* abbr. com., comm. **1.** The act of communicating; transmission. **2. a.** The exchange of thoughts, messages, or information, as by speech, signals, writing, or behavior. **b.** Interpersonal rapport. **c.** communications: (used with a sing. or pl. verb) The art and technique of using words effectively and with grace in imparting one's ideas.

This "exchange of thoughts, messages, or information" can take place using a number of different media and technologies.

Labels

Regardless of how you might feel about Occupational Safety and Health Administration requirements, the application of OSHA standards, or the policies of the state or federal agencies responsible for monitoring compliance, the reasoning behind the legislation is sound. Everyone should have the right to know whenever and wherever they may be exposed to a material, chemical, or biological hazard, or when their personal safety may be compromised.

It is the responsibility of the business owner to identify and communicate the nature of these hazards to anyone and everyone exposed to such hazards, and that includes clients and customers as well as employees. However, it is the responsibility of each employee (with the active participation of management) to be aware of chemical dangers in the workplace. The implication is clear: facilitating such awareness requires an ongoing education and training program.

In addition to training, however, all chemical containers must be labeled with the chemical name, its general function for us, and the dangers and hazards associated with its use and handling, and that label must remain clearly written and legible at all times.

Signs and Posters

There are also very specific requirements regarding signs, posters, and the identification of known safety or chemical hazards. Some of these signs can be meant solely for the purpose of reminding an employee to exercise care in the execution of certain job related tasks, such as connecting a ground cable after connecting the positive cable and removing it first when jumping or "boosting" a battery, or they can be used to identify the location of personal protective equipment, various locations of first aid kits, or where clean up materials are stored. Others, posters in particular, are meant to clearly identify the nature of the employee's right to know and the particulars of the applicable legislation.

Material Safety Data Sheets

OSHA legislation requires the manufacturers of hazardous materials to compile information specifically related to the nature and use of such chemicals in the workplace and then make that information available in the form of a Material Safety Data Sheet, or as it is commonly known in industry, an MSDS. This information must meet certain government guidelines that have been designed to ensure at least some degree of consistency (this is sometimes hard to recognize) and ease of use as a reference, especially in the case of an accident or an emergency. Every chemical that can or may provide a biological, chemical, or potential safety hazard in the workplace should have the MSDS associated with it on file and accessible within the facility. If it doesn't, request it from your suppliers. If they fail to provide it, look for the material on the Internet. And if it is not available there, contact the nearest office of the Occupational Safety and Health Administration.

The MSDS should include the name of the substance; the manufacturer, including contact information; and when the MSDS document itself was last prepared. It should also include any other names by which that particular chemical is known and its exposure limits. Exposure limits are generally defined in three different ways, especially with regard to health, safety, and the workplace. Consistent with the alphabet soup of acronyms embraced by most government agencies, these limits are classified as OELs (Occupational Exposure Limits), MELs (Maximum Exposure Limits), and PELs (Permissible Exposure Limits). All three have to do with the maximum amount of a compound or chemical to which an individual may be *safely* exposed in the course of a normal work day without adverse effect. This is usually calculated using the volume of the chemical in question and the duration of the exposure.

The physical characteristics of the chemical are documented in the MSDS, which can include what the chemical looks like, what it smells like, the point at which it is likely to melt or boil, if it is likely to dissolve and in what, and its buoyancy. Whether or not the chemical or substance is flammable or explosive, and the circumstances under which either is likely or unlikely to occur, should also be identified on the MSDS. This should include the best way to extinguish a fire should one occur, and the nature of any vapors that might result from combustion that is the result of fire or explosion.

The Material Safety and Data Sheet should also discuss what is known as reactivity—how that chemical or substance is likely to react with other chemicals or substances, including compounds as simple and basic as air and water.

The health hazards section of an MSDS should detail how the chemical or compound might enter the body, such as inhalation, absorption through contact with the skin, ingestion, or injection; while the section dealing with handling and storage should describe how to clean up a spill or leak in addition to the proper environment for storage, transportation, and/or disposal. Finally, the MSDS should identify what personal protective gear or equipment should be associated with the handling, storage, use, and disposal of that particular chemical or hazard—such as respirators, goggles, ear protection, gloves, aprons, etc.

Included here is an example of a Material Safety Data Sheet for WD-40, a popular penetrating oil and lubricant. As you can see from the illustrations in Figures 1-2 and 1-3, just about anything you might want or need to know about the safe handling of this product is included within these two pages of information.

WD-40

MATERIAL SAFETY DATA SHEET

I. PRODUCT IDENTIFICATION

Manufacturer:	WD-40 Company	Telephone:	
Address:	1061 Cudahy Place (92110) P.O. Box 80607 San Diego, California 92138-0607	Emergency Only:	1 (800) 424-9300 (CHEMTREC)
		Information:	(619) 275-1400
		Chemical Name:	Organic Mixture
		Trade Name:	WD-40 Bulk Liquid

II. HAZARDOUS INGREDIENTS

Chemical Name	CAS Number	%	Exposure Limit ACGIH/OSHA
Aliphatic Petroleum Distillates	8052-41-3	70	100 ppm PEL
Petroleum Base Oil	64742-65-0	> 20	5 mg/M³ TWA (mist)
Non-hazardous Ingredients		< 10	

III. PHYSICAL DATA

Boiling Point:	300°F (minimum)	Evaporation Rate:	Not determined
Vapor Density (air = 1):	Greater than 1	Vapor Pressure:	Not determined
Solubility in Water:	Insoluble	Appearance:	Cloudy light amber
Specific Gravity (H_2O = 1):	0.816 @ 70°F	Odor:	Characteristic odor
Percent Volatile (volume):	70%	VOC:	566 grams per liter

IV. FIRE AND EXPLOSION

Flash Point:	Tag Open Cup 110°F (minimum)
Flammable Limits:	(Solvent Portion) [Lel] 1.0% [Uel] 6.0%
Extinguishing Media:	CO_2, Dry Chemical, Foam
Special Fire Fighting Procedures:	None
Unusual Fire and Explosion Hazards:	None

V. HEALTH HAZARD / ROUTE(S) OF ENTRY

Threshold Limit Value

Aliphatic Petroleum Distillates (Stoddard Solvent) lowest TLV (ACGIH 100 ppm.)

Symptoms of Overexposure

Inhalation (Breathing):	May cause anesthesia, headache, dizziness, nausea and upper respiratory irritation.
Skin Contact:	May cause drying of skin and/or irritation.
Eye Contact:	May cause irritation, tearing and redness.
Ingestion (Swallowed):	May cause irritation, nausea, vomiting and diarrhea.

First Aid Emergency Procedures

Ingestion (Swallowed):	Do not induce vomiting, seek medical attention.
Eye Contact:	Immediately flush eyes with large amounts of water for 15 minutes.
Skin Contact:	Wash with soap and water.
Inhalation (Breathing):	Remove to fresh air. Give artificial respiration if necessary. If breathing is difficult, give oxygen.

Pre-existing medical conditions such as eye, skin and respiratory disorders may be aggravated by exposure.

DANGER!

Aspiration Hazard: If swallowed, can enter lungs and may cause chemical pneumonitis. Do not induce vomiting. Call Physician immediately.

Suspected Cancer Agent

Yes ______ No __X__ — The components in this mixture have been found to be noncarcinogenic by NTP, IARC and OSHA.

Figure 1-2 WD40 MSDS Items I–V.

VI. REACTIVITY DATA

Stability:	Stable __X__	Unstable ____
Conditions to avoid:	NA	
Incompatibility:	Strong oxidizing materials	
Hazardous decomposition products:	Thermal decomposition may yield carbon monoxide and/or carbon dioxide.	
Hazardous polymerization:	May occur ____	Will not occur __X__

VII. SPILL OR LEAK PROCEDURES

Spill Response Procedures
Absorb small quantities with sand, earth, sawdust. Large quantities pump into tank.
Waste Disposal Method
Incinerate liquid, bury saturated absorbent in land fill. Dispose of in accordance with local, state and federal regulations.

VIII. SPECIAL HANDLING INFORMATION

Ventilation:	Sufficient to keep solvent vapor less than TLV.
Respiratory Protection:	Advised when concentrations exceed TLV.
Protective Gloves:	Advised to prevent possible skin irritation.
Eye Protection:	Approved eye protection to safeguard against potential eye contact, irritation or injury.
Other Protective Equipment:	None required.

IX. SPECIAL PRECAUTIONS

Keep from open flame, do not take internally. Avoid excessive inhalation of spray particles. Keep from children.

X. TRANSPORTATION DATA (49 CFR 172.101)

Domestic Surface	
Description:	WD-40 Bulk Lubricant
Hazard Class:	Non-Regulated Per 49 CFR 173.150(F)(2)
ID No.:	None
Label Required:	None (Under 119 Gallons)
Domestic Air	
Description:	Petroleum Distillates N.O.S.
Hazard Class:	3 UN 1268 PG III
Label Required:	Flammable Liquid

XI. REGULATORY INFORMATION

All ingredients for this product are listed on the TSCA inventory.

SARA Title III chemicals:	None
California Prop 65 chemicals:	None
CERCLA reportable quantity:	None
RCRA hazardous waste no.:	D001 (Ignitable)

SIGNATURE: R. Miles TITLE: Technical Director
REVISION DATE: March 2001 SUPERSEDES: March 1998

NA = Not applicable NDA = No data available < = Less than > = More than

We believe the statements, technical information and recommendations contained herein are reliable. However, the data is provided without warranty, expressed or implied. It is the user's responsibility both to determine safe conditions for use of this product and assume loss, damage or expense, direct or consequential, arising from its use. Before using product, read label.

MSDS-6

Figure 1-3 WD40 MSDS Items VI–XI.

Manuals and Manifests

In any dealings with a hazardous material, you would be prudent to assume that you are responsible for that material from cradle to crypt. In other words, if you accepted it from the vehicle owner, had anything to do with removing it from the vehicle, allowed it drip onto your driveway or floor or escape into the atmosphere, you are the responsible party the government will come looking for if there is a problem. Even if you feel you are merely acting as a conduit for that material as it moves from its actual source—the vehicle and the vehicle owner—to the hazardous material disposal site, because it was only in your care and custody for a very short period of time, it's still your baby. Consequently, it would be wise to attack the record-keeping requirements of both the Environmental Protection Agency (state or federal) and the Occupational Safety and Health Administration aggressively in order to ensure a paper trail that is squeaky clean. It would almost be safe to translate the above as: DO NOT EVER THROW OUT ANY DOCUMENT RELATED TO HAZARDOUS MATERIALS MANAGEMENT, MAINTENANCE, or DISPOSAL IF YOU CAN AVOID IT!!! You just may need it to prove that you did what you said you did, and not having it might be the difference between years of litigation, costly settlements, bankruptcy, or worse. Keeping these records forever, or for at least as long as you can regardless of what the law requires, may be the least expensive way to protect yourself if your actions are ever called into question by a government agency or the courts.

In addition, it's important to know whom you are dealing with. And it wouldn't hurt to know with whom they are dealing either. Be sure you are dealing with a management or disposal company that is licensed, bonded, fully insured, experienced, reasonable, responsible, and perhaps most important of all, substantial. It does not hurt to be involved with a company that is an exponentially bigger fish with deeper pockets than you are, especially in the eyes of a litigator, or a government agent. That advice might not appear terribly altruistic, and maybe it's not. But it may save your life and/or your business someday!

DOCUMENTATION, RECORDS, AND TRAINING

It is not good enough to institute a superficial training program designed to simply get OSHA, state or federal, "off your back." The Process Safety Management Standard requires the employer "to evaluate or verify that employees comprehend the training given to them." This implies that the training offered must meet certain requirements, the first of which is clearly defined goals and objects. To a large degree, this means the training result must be both measurable and measured. And because this is an OSHA requirement, documenting the nature of the training, its goals and objectives, when and where the training took place, and of course its results is more than just a good idea—it is the law and, consequently, essential. In fact, according to an OSHA booklet entitled, *Training Requirements in Osha Standards and Training Guidelines*, US Department of Labor, Occupational Safety and Health Administration, OSHA 2254 1998 (Revised), documentation "can provide evidence of the employer's good faith and compliance with OSHA standards." This kind of documentation can also answer one of the first questions an accident investigator is likely to ask: Was the injured employee trained to do the job he was being asked to do?

You can find the guidelines for designing, conducting, establishing, evaluating, and then continually improving this kind of training in the above mentioned OSHA publication. However, documentation is incomplete without proper record-keeping and just having these records "around here someplace" will not be enough.

Remember, there may come a time when you will actually need to get your hands on the paperwork that serves as a living record of your active and voluntary compliance with these standards.

PERSONAL PROTECTIVE EQUIPMENT

It is virtually impossible to work in the automotive service environment and at the same time insulate yourself from all potential dangers and hazards that exist there. There are laundry lists that can be created from the carcinogenic chemicals used in the everyday performance of our profession alone. Exposure to the wrong chemical cleaning agents or even exposure to the *right* chemical cleaning agents for extended periods of time can and often do have serious side effects. Battery electrolyte is not much more than sulfuric acid and water, and anyone who is old enough to have actually *filled* a battery and gotten some battery acid in an open cut knows firsthand that it can't be too good for you if it can hurt that much!

Most shops purchase brake cleaning chemicals by the caseload, and there are times when you can smell a brake job long before you can actually see it being performed. Carburetor cleaners are capable of doing much more than cleaning carburetors, as evidenced by the way they can dissolve the hair on your hands and arms. Sparks fly in all directions when surfaces are being cut or cleaned with high speed tools that utilize rotating discs made of abrasive materials, and the noise level that results from air tools alone is quite literally deafening.

When I grew up in the industry no one thought twice about using oxygen and acetylene to cut or weld, with or without the use of protective goggles. No one thought twice about sticking his hand or his arm elbow deep into chemicals the very nature of which can cause nightmares today. Oil and grease and any other toxic chemical waste was generally cleaned or mopped up with the polluted or contaminated result poured down the drain, if not just dumped out on to the street. It seems somehow inadequate to insist that no one knew any better. But realistically, it was the truth. We didn't! Or at least, we didn't think about it!

However, we do know better today, and there is no excuse—at least, no reasonable excuse—to place yourself or anyone else in harm's way because you don't have the proper gear to do the job quickly and safely. That means eye protection, ear protection, the proper gloves, steel-toed boots if necessary, and protective aprons if wearing them is warranted.

We'll talk more about personal protective equipment (PPE) later on. But for now, here at the introduction to *Safety Communications*, let's just say that anything less than extreme caution is criminal. Why? Because, in fact, it is!

MATERIAL HANDLING AND STORAGE

"A place for everything and everything in its place!"

— *Your Mother*

There is a lot of truth in the old saying that there should be "a place for everything, and everything in its place!" And more truth, perhaps, to the old saying that suggests there is a "right way and a wrong way to do everything!" Nowhere is this more true than when it comes to handling and storing hazardous or potentially hazardous materials.

It is critical that potentially dangerous chemicals, gases, liquids, or powders are stored in areas where they are least likely to present a hazard. Things that adversely react when exposed to extreme heat should be stored where it is always cool or cold. Materials that don't do well when frozen should be kept someplace where the temperature is more moderate. To prevent chemical spills, many states have laws that require the use of double-walled containment tanks that are capable of holding a minimum of one-and-a-half times the capacity of the storage tank or more, before exposing the liquid being stored to the environment. There are absorbents to contain spills if and when they might occur, and standards and procedures for disposing of the absorbents when the time comes.

SUMMARY

Our capacity to learn and grow is better understood within the context of the four stages of knowledge. It is within these four stages that you can clearly see the path to self-awareness and self-actualization blossom. This journey begins with the realization that you don't know what you don't know and continues through the development of what the Zen masters refer to as "the beginner's mind." This was perhaps best explained by Bruce Lee, the famous martial artist, in an article that appeared in *Black Belt Magazine*, September 1971, in which Lee expanded upon an ancient Zen parable.

A learned man once went to a Zen teacher to inquire about Zen. As the Zen teacher explained, the learned man would frequently interrupt him with remarks like, "Oh, yes, we have that too...." and so on.

Finally the Zen teacher stopped talking and began to serve tea to the learned man. He poured the cup full, and then kept pouring until the cup overflowed.

"Enough!" the learned man once more interrupted. "No more can go into the cup!"

"Indeed, I see," answered the Zen teacher. "If you do not first empty the cup, how can you taste my cup of tea?"

The journey to wisdom begins with an empty cup and a "beginner's mind": the realization that you don't know what you don't know. And that in order to grow, you must know/recognize/identify what you need to know: the second of the four stages. Stage 2 is the beginning of wisdom because it seeks to isolate what we need to learn in order to be successful. Stage 3 occurs when we realize that through exposure to new ideas, skills, and personal experiences we are able to see that we don't always know what we actually know! Stage 4 is the stage in which we come to realize that we know what we know, and from that point the journey begins again as we enter a new and unfamiliar area of performance.

The world of safety communications and hazardous material responsibilities is one area in which this journey through the four stages of knowledge becomes eminently clear.

It would be safe to say that most shop owners around the country have little or no idea what their hazmat or safety communication responsibilities are with regard to their employees, their customers, or the regulatory agencies that will ultimately determine whether or not those responsibilities have been discharged adequately—they don't know what they don't know.

In Chapter 1, you find an overview of exactly what those responsibilities are, along with the other side of the *owner's responsibility coin*, the side that deals with *employee rights*: the right to know, the right to training, the right to information, the right to participate and demand action, and the right of access. Each of these rights brings with it a specific set of responsibilities that you, as an owner or manager, must be prepared to discharge quickly, completely, and most of all, effectively. These responsibilities include that of complying with both the letter and the spirit of the law, especially with regard to hazardous materials communication.

This includes your responsibility as an owner or manager to see that your shop is properly audited for safety and material hazards; that all hazards have been properly identified and labeled; that all appropriate signs and posters are in place; that there are Material Safety Data Sheets available for each hazardous material stored on-site; and that all of the above has been properly recorded and documented, including your on-going hazardous materials training efforts. It also mandates that the appropriate personal protective equipment for each task is identified and available and that all material hazards are handled and stored properly.

If Chapter 1 is consumed with reconciling what we don't know with what we need to know, the balance of the book will be focused on the remaining stages of knowledge until we reach Stage 4, the point at which we know what we know and can deal with all of our safety communication responsibilities confidently and completely.

Write down some of the areas you feel you will need to concentrate your efforts in order to be successful in the chart provided on the following page.

Thoughts

Actions

Results

CHAPTER 2

Management Responsibilities

INTRODUCTION

This chapter introduces you to the safety communications responsibilities faced by a business owner or manager, with specific regard to:

- Safety audit and facility assessment
- Safety training
- Employee right to know
- Safety equipment and procedures
- Meetings

SAFETY AUDIT AND FACILITY ASSESSMENT

In Chapter 1, we presented a fairly substantial laundry list of shop or employer responsibilities that is almost overwhelming. Unfortunately, those responsibilities are required by law and therefore necessary, no matter how intimidating they might seem. Realistically, the best way to ensure that you are in compliance is by conducting an audit of your workplace with special attention to each of these areas. To facilitate this exercise, OSHA has actually created a number of checklists and training aids specifically designed to help you as a small business owner. Among the best examples of these are the Self-Inspection Checklists for general industry that were created in 1997 and made available on OSHA's website at: *osha.gov/SLTC/smallbusiness/chklist.html*

As you can see, this self-auditing tool covers almost every area of compliance that you must be concerned with as a small business owner or manager. For your consideration, I have provided the checklist in its entirety in order to allow you to see the specifics of what a government auditor might be looking for.

SAFETY AND HEALTH PROGRAM

- ❑ Do you have an active safety and health program in operation that deals with general safety and health program elements as well as management of hazards specific to your worksite?
- ❑ Is one person clearly responsible for the overall activities of the safety and health program?
- ❑ Do you have a safety committee or group made up of management and labor representatives that meets regularly and reports in writing on its activities?
- ❑ Do you have a working procedure for handling in-house employee complaints regarding safety and health?
- ❑ Are you keeping your employees advised of the successful effort and accomplishments you and/or your safety committee have made in assuring that they will have a workplace that is safe and healthful?
- ❑ Have you considered incentives for employees or workgroups who have excelled in reducing workplace injuries/illnesses?

PERSONAL PROTECTIVE EQUIPMENT (PPE)

- ❑ Are employers assessing the workplace to determine if hazards that require the use of personal protective equipment (for example, head, eye, face, hand, or foot protection) are present or are likely to be present?
- ❑ If hazards or the likelihood of hazards are found, are employers selecting, and having affected employees use, properly fitted personal protective equipment suitable for protection from these hazards?
- ❑ Has the employee been trained on PPE procedures, that is, what PPE is necessary for a job task, when it is needed, and how to properly adjust it?
- ❑ Are protective goggles or face shields provided and worn where there is any danger of flying particles or corrosive materials?

  Are approved safety glasses required to be worn at all times in areas where there is a risk of eye injuries such as punctures, abrasions, contusions, or burns?

- ❑ Are employees who need corrective lenses (glasses or contacts) in working environments having harmful exposures required to wear only approved safety glasses or protective goggles, or use other medically approved precautionary procedures?
- ❑ Are protective gloves, aprons, shields, or other means provided and required where employees could be cut or where there is reasonably anticipated exposure to corrosive liquids, chemicals, blood, or other potentially infectious materials (see 29 CFR 1910.1030(b) for the definition of "other potentially infectious materials")?
- ❑ Are hard hats provided and worn where danger of falling objects exists?
- ❑ Are hard hats inspected periodically for damage to the shell and suspension system?
- ❑ Is appropriate foot protection required where there is the risk of foot injuries from hot, corrosive, or poisonous substances, falling objects, crushing or penetrating actions?
- ❑ Are approved respirators provided for regular or emergency use where needed?
- ❑ Is all protective equipment maintained in a sanitary condition and ready for use?
- ❑ Do you have eye wash facilities and a quick drench shower within the work area where employees are exposed to injurious corrosive materials? Where special equipment is needed for electrical workers, is it available?
- ❑ Where food or beverages are consumed on the premises, are they consumed in areas where there is no exposure to toxic material, blood, or other potentially infectious materials?
- ❑ Is protection against the effects of occupational noise exposure provided when sound levels exceed those of the OSHA noise standard?
- ❑ Are adequate work procedures, protective clothing and equipment provided and used when cleaning up spilled toxic or otherwise hazardous materials or liquids?
- ❑ Are there appropriate procedures in place for disposing of or decontaminating personal protective equipment contaminated with, or reasonably anticipated to be contaminated with, blood or other potentially infectious materials?

FLAMMABLE AND COMBUSTIBLE MATERIALS

- ❑ Are combustible scrap, debris, and waste materials (oily rags, etc.) stored in covered metal receptacles and removed from the worksite promptly?
- ❑ Is proper storage practiced to minimize the risk of fire, including spontaneous combustion?
- ❑ Are approved containers and tanks used for the storage and handling of flammable and combustible liquids?
- ❑ Are all connections on drums and combustible liquid piping, vapor and liquid tight?
- ❑ Are all flammable liquids kept in closed containers when not in use (for example, parts cleaning tanks, pans, etc.)?
- ❑ Are bulk drums of flammable liquids grounded and bonded to containers during dispensing?
- ❑ Do storage rooms for flammable and combustible liquids have explosion-proof lights?
- ❑ Do storage rooms for flammable and combustible liquids have mechanical or gravity ventilation?

- ❑ Is liquefied petroleum gas stored, handled, and used in accordance with safe practices and standards?
- ❑ Are "NO SMOKING" signs posted on liquefied petroleum gas tanks?
- ❑ Are liquefied petroleum storage tanks guarded to prevent damage from vehicles?
- ❑ Are all solvent wastes and flammable liquids kept in fire-resistant, covered containers until they are removed from the worksite?
- ❑ Is vacuuming used whenever possible rather than blowing or sweeping combustible dust? Are firm separators placed between containers of combustibles or flammables, when stacked one upon another, to assure their support and stability?
- ❑ Are fuel gas cylinders and oxygen cylinders separated by distance, and fire-resistant barriers, while in storage?
- ❑ Are fire extinguishers selected and provided for the types of materials in areas where they are to be used?
    - ❑ Class A: Ordinary combustible material fires.
    - ❑ Class B: Flammable liquid, gas or grease fires.
    - ❑ Class C: Energized-electrical equipment fires.
- ❑ Are appropriate fire extinguishers mounted within 75 feet of outside areas containing flammable liquids, and within 10 feet of any inside storage area for such materials?
- ❑ Are extinguishers free from obstructions or blockage?
- ❑ Are all extinguishers serviced, maintained, and tagged at intervals not to exceed one year?
- ❑ Are all extinguishers fully charged and in their designated places?
- ❑ Where sprinkler systems are permanently installed, are the nozzle heads so directed or arranged that water will not be sprayed into operating electrical switch boards and equipment?
- ❑ Are "NO SMOKING" signs posted where appropriate in areas where flammable or combustible materials are used or stored?
- ❑ Are safety cans used for dispensing flammable or combustible liquids at a point of use?
- ❑ Are all spills of flammable or combustible liquids cleaned up promptly?
- ❑ Are storage tanks adequately vented to prevent the development of excessive vacuum or pressure as a result of filling, emptying, or atmosphere temperature changes?
- ❑ Are storage tanks equipped with emergency venting that will relieve excessive internal pressure caused by fire exposure?
- ❑ Are "NO SMOKING" rules enforced in areas involving storage and use of hazardous materials?

HAND AND PORTABLE-POWERED TOOLS

- ❑ Are all tools and equipment (both company and employee owned) used by employees at their workplace in good condition?
- ❑ Are hand tools such as chisels and punches, which develop mushroomed heads during use, reconditioned or replaced as necessary?

- ❑ Are broken or fractured handles on hammers, axes, and similar equipment replaced promptly?
- ❑ Are worn or bent wrenches replaced regularly?
- ❑ Are appropriate handles used on files and similar tools?
- ❑ Are employees made aware of the hazards caused by faulty or improperly used hand tools?
- ❑ Are appropriate safety glasses, face shields, etc., used with hand tools or equipment which might produce flying materials or be subject to breakage?
- ❑ Are jacks checked periodically to ensure that they are in good operating condition?
- ❑ Are tool handles wedged tightly in the head of all tools?
- ❑ Are tool cutting edges kept sharp so the tool will move smoothly without binding or skipping?
- ❑ Are tools stored in dry, secure locations where they won't be tampered with?
- ❑ Is eye and face protection used when driving hardened or tempered spuds or nails?

Portable (Power-Operated) Tools and Equipment

- ❑ Are grinders, saws, and similar equipment provided with appropriate safety guards?
- ❑ Are power tools used with the correct shield, guard, or attachment, recommended by the manufacturer?
- ❑ Are portable circular saws equipped with guards above and below the base shoe? Are circular saw guards checked to ensure that they are not wedged up, thus leaving the lower portion of the blade unguarded?
- ❑ Are rotating or moving parts of equipment guarded to prevent physical contact?
- ❑ Are all cord-connected, electrically operated tools and equipment effectively grounded or of the approved double insulated type?
- ❑ Are effective guards in place over belts, pulleys, chains, sprockets, on equipment such as concrete mixers and air compressors?
- ❑ Are portable fans provided with full guards or screens having openings 1/2 inch or less?
- ❑ Is hoisting equipment available and used for lifting heavy objects, and are hoist ratings and characteristics appropriate for the task?
- ❑ Are ground fault circuit interrupters provided on all temporary electrical 15- and 20-ampere circuits used during periods of construction?
- ❑ Are pneumatic and hydraulic hoses on power operated tools checked regularly for deterioration or damage?

Powder-Actuated Tools

- ❑ Are employees who operate powder-actuated tools trained in their use and do they carry a valid operator's card?
- ❑ Is each powder-actuated tool stored in its own locked container when not being used?
- ❑ Is a sign at least seven inches by ten inches with bold face type reading "POWDER-ACTUATED TOOL IN USE" conspicuously posted when the tool is being used?

- ❑ Are powder-actuated tools left unloaded until they are actually ready to be used?
- ❑ Are powder-actuated tools inspected for obstructions or defects each day before use?
- ❑ Do powder-actuated tool operators have and use appropriate personal protective equipment such as hard hats, safety goggles, safety shoes, and ear protectors?

LOCKOUT/TAGOUT PROCEDURES

- ❑ Is all machinery or equipment which is capable of movement required to be de-energized or disengaged and locked out during cleaning, servicing, adjusting, or setting up operations?

  Where the power disconnecting method for equipment does not also disconnect the electrical control circuit:

- ❑ Are the appropriate electrical enclosures identified?
- ❑ Are means provided to ensure that the control circuit can also be disconnected and locked out?
- ❑ Is the locking out of control circuits in lieu of locking out main power disconnects prohibited?
- ❑ Are all equipment control valve handles provided with a means for locking out?
- ❑ Does the lockout procedure require that stored energy (mechanical, hydraulic, air, etc.) be released or blocked before equipment is locked out for repairs?
- ❑ Are appropriate employees provided with individually keyed personal safety locks?
- ❑ Are employees required to keep personal control of their key(s) while they have safety locks in use?
- ❑ Is it required that only the employee exposed to the hazard may place or remove the safety lock?
- ❑ Is it required that employees check the safety of the lockout by attempting a startup after making sure no one is exposed?
- ❑ Are employees instructed to always push the control circuit stop button immediately after checking the safety of the lockout?
- ❑ Is there a means provided to identify any or all employees who are working on locked out equipment by their locks or accompanying tags?
- ❑ Are a sufficient number of accident preventive signs or tags and safety padlocks provided for any reasonably foreseeable repair emergency?
- ❑ When machine operations, configuration, or size requires the operator to leave his or her control station to install tools or perform other operations, and that part of the machine could move if accidentally activated, is such element required to be separately locked or blocked out?
- ❑ In the event that equipment or lines cannot be shut down, locked out, and tagged, is a safe job procedure established and rigidly followed?

CONFINED SPACES

- ❑ Are confined spaces thoroughly emptied of any corrosive or hazardous substances, such as acids or caustics, before entry?
- ❑ Are all lines to a confined space containing inert, toxic, flammable, or corrosive materials valved off and blanked or disconnected and separated before entry?

- ❑ Are all impellers, agitators, or other moving parts and equipment inside confined spaces locked out if they present a hazard?
- ❑ Is either natural or mechanical ventilation provided prior to confined space entry?
- ❑ Are appropriate atmospheric tests performed to check for oxygen deficiency, toxic substances, and explosive concentrations in the confined space before entry?
- ❑ Is adequate illumination provided for the work to be performed in the confined space?
- ❑ Is the atmosphere inside the confined space frequently tested or continuously monitored during conduct of work? Is there an assigned safety standby employee outside the confined space, when required, whose sole responsibility is to watch the work in progress, sound an alarm if necessary, and render assistance?
- ❑ Is the standby employee appropriately trained and equipped to handle an emergency?
- ❑ Is the standby employee or other employees prohibited from entering the confined space without lifelines and respiratory equipment if there is any question as to the cause of an emergency?
- ❑ Is approved respiratory equipment required if the atmosphere inside the confined space cannot be made acceptable?
- ❑ Is all portable electrical equipment used inside confined spaces either grounded and insulated, or equipped with ground fault protection?
- ❑ Before gas welding or burning is started in a confined space, are hoses checked for leaks, compressed gas bottles forbidden inside of the confined space, torches lighted only outside of the confined area, and the confined area tested for an explosive atmosphere each time before a lighted torch is to be taken into the confined space?
- ❑ If employees will be using oxygen-consuming equipment such as salamanders, torches, and furnaces in a confined space, is sufficient air provided to assure combustion without reducing the oxygen concentration of the atmosphere below 19.5 percent by volume?
- ❑ Whenever combustion-type equipment is used in a confined space, are provisions made to ensure that the exhaust gases are vented outside the enclosure?
- ❑ Is each confined space checked for decaying vegetation or animal matter which may produce methane?
- ❑ Is the confined space checked for possible industrial waste which could contain toxic properties?
- ❑ If the confined space is below the ground and near areas where motor vehicles will be operating, is it possible for vehicle exhaust or carbon monoxide to enter the space?

ELECTRICAL

- ❑ Do you specify compliance with OSHA for all contract electrical work?
- ❑ Are all employees required to report as soon as practicable any obvious hazard to life or property observed in connection with electrical equipment or lines?

- ❑ Are employees instructed to make preliminary inspections and/or appropriate tests to determine what conditions exist before starting work on electrical equipment or lines?
- ❑ When electrical equipment or lines are to be serviced, maintained, or adjusted, are necessary switches opened, locked out, and tagged whenever possible?
- ❑ Are portable electrical tools and equipment grounded or of the double insulated type?
- ❑ Are electrical appliances such as vacuum cleaners, polishers, and vending machines grounded?
- ❑ Do extension cords in use have a grounding conductor?
- ❑ Are multiple plug adaptors prohibited?
- ❑ Are ground fault circuit interrupters installed on each temporary 15- or 20-ampere, 120-volt AC circuit at locations where construction, demolition, modifications, alterations, or excavations are being performed?
- ❑ Are all temporary circuits protected by suitable disconnecting switches or plug connectors at the junction with permanent wiring?
- ❑ Do you have electrical installations in hazardous dust or vapor areas? If so, do they meet the National Electrical Code (NEC) for hazardous locations?
- ❑ Are exposed wiring and cords with frayed or deteriorated insulation repaired or replaced promptly?
- ❑ Are flexible cords and cables free of splices or taps?
- ❑ Are clamps or other securing means provided on flexible cords or cables at plugs, receptacles, tools, equipment, etc., and is the cord jacket securely held in place? Are all cord, cable, and raceway connections intact and secure?
- ❑ In wet or damp locations, are electrical tools and equipment appropriate for the use or location, or otherwise protected?
- ❑ Is the location of electrical power lines and cables (overhead, underground, underfloor, other side of walls) determined before digging, drilling, or similar work is begun?
- ❑ Are metal measuring tapes, ropes, handlines, or similar devices with metallic thread woven into the fabric prohibited where they could come in contact with energized parts of equipment or circuit conductors?
- ❑ Is the use of metal ladders prohibited in areas where the ladder or the person using the ladder could come in contact with energized parts of equipment, fixtures, or circuit conductors?
- ❑ Are all disconnecting switches and circuit breakers labeled to indicate their use or the equipment served?
- ❑ Are disconnecting means always opened before fuses are replaced?
- ❑ Do all interior wiring systems include provisions for grounding metal parts of electrical raceways, equipment, and enclosures?
- ❑ Are all electrical raceways and enclosures securely fastened in place?
- ❑ Are all energized parts of electrical circuits and equipment guarded by approved cabinets or enclosures against accidental contact?

- ❑ Is sufficient access and working space provided and maintained around all electrical equipment to permit ready and safe operations and maintenance?
- ❑ Are all unused openings (including conduit knockouts) in electrical enclosures and fittings closed with appropriate covers, plugs, or plates?
- ❑ Are electrical enclosures such as switches, receptacles, and junction boxes provided with tight fitting covers or plates?
- ❑ Are disconnecting switches for electrical motors in excess of two horsepower capable of opening the circuit when the motor is in a stalled condition, without exploding? (Switches must be horsepower rated equal to or in excess of the motor hp rating.) Is low voltage protection provided in the control device of motors driving machines or equipment which could cause probable injury from inadvertent starting?
- ❑ Is each motor disconnecting switch or circuit breaker located within sight of the motor control device?
- ❑ Is each motor located within sight of its controller or the controller disconnecting means capable of being locked in the open position, or is a separate disconnecting means installed in the circuit within sight of the motor?
- ❑ Is the controller for each motor in excess of two horsepower rated in horsepower equal to or in excess of the rating of the motor it serves?
- ❑ Are employees who regularly work on or around energized electrical equipment or lines instructed in cardiopulmonary resuscitation (CPR) methods?
- ❑ Are employees prohibited from working alone on energized lines or equipment over 600 volts?

WALKING/WORKING SURFACES

General Work Environment

- ❑ Is a documented, functioning housekeeping program in place?
- ❑ Are all worksites clean, sanitary, and orderly?
- ❑ Are work surfaces kept dry, or are appropriate means taken to ensure that the surfaces are slip-resistant?
- ❑ Are all spilled hazardous materials or liquids, including blood and other potentially infectious materials, cleaned up immediately and according to proper procedures?
- ❑ Is combustible scrap, debris, and waste stored safely and removed from the worksite properly?
- ❑ Is all regulated waste, as defined in the OSHA blood-borne pathogens standard (1910.1030), discarded according to federal, state, and local regulations?
- ❑ Are accumulations of combustible dust routinely removed from elevated surfaces, including the overhead structure of buildings, etc.?
- ❑ Is combustible dust cleaned up with a vacuum system to prevent the dust from going into suspension?
- ❑ Is metallic or conductive dust prevented from entering or accumulating on or around electrical enclosures or equipment?

- ❑ Are covered metal waste cans used for oily and paint-soaked waste?

Walkways

- ❑ Are aisles and passageways kept clear?
- ❑ Are aisles and walkways marked as appropriate?
- ❑ Are wet surfaces covered with non-slip materials?
- ❑ Are holes in the floor, sidewalk, or other walking surface repaired properly, covered, or otherwise made safe?
- ❑ Is there safe clearance for walking in aisles where motorized or mechanical handling equipment is operating?
- ❑ Are materials or equipment stored in such a way that sharp projections will not interfere with the walkway?
- ❑ Are spilled materials cleaned up immediately?
- ❑ Are changes of direction or elevation readily identifiable?
- ❑ Are aisles or walkways that pass near moving or operating machinery, welding operations, or similar operations arranged so employees will not be subjected to potential hazards?
- ❑ Is adequate headroom provided for the entire length of any aisle or walkway?
- ❑ Are standard guardrails provided wherever aisle or walkway surfaces are elevated more than thirty inches above any adjacent floor or the ground?
- ❑ Are bridges provided over conveyors and similar hazards?

Floor and Wall Openings

- ❑ Are floor openings guarded by a cover, a guardrail, or equivalent device on all sides (except at entrance to stairways or ladders)?
- ❑ Are toeboards installed around the edges of permanent floor openings (where persons may pass below the opening)?
- ❑ Are skylight screens of such construction and mounting that they will withstand a load of at least 200 pounds?
- ❑ Is the glass in the windows, doors, glass walls, etc., which are subject to human impact, of sufficient thickness and type for the condition of use?
- ❑ Are grates or similar type covers over floor openings, such as floor drains, of such design that foot traffic or rolling equipment will not be affected by the grate spacing?
- ❑ Are unused portions of service pits and pits not actually in use either covered or protected by guardrails or the equivalent?
- ❑ Are manhole covers, trench covers, and similar covers, plus their supports, designed to carry a truck rear axle load of at least 20,000 pounds when located in roadways and subject to vehicle traffic?
- ❑ Are floor or wall openings in fire-resistive structures provided with doors or covers compatible with the fire rating of the structure and provided with a self-closing feature when appropriate?

Stairs and Stairways

- ❑ Are there standard stair rails or handrails on all stairways having four or more risers?
- ❑ Are all stairways at least twenty-two inches wide?
- ❑ Do stairs have landing platforms measuring not less than thirty inches in the direction of travel and twenty-two inches in width at every twelve feet or less of vertical rise?
- ❑ Do stairs angle no more than fifty and no less than thirty degrees?
- ❑ Are step risers on stairs uniform from top to bottom?
- ❑ Are steps on stairs and stairways designed or provided with a surface that renders them slip-resistant?
- ❑ Are stairway handrails located between thirty and thirty-four inches above the leading edge of stair treads?
- ❑ Do stairway handrails have at least three inches of clearance between the handrails and the wall or surface they are mounted on?
- ❑ Where doors or gates open directly on a stairway, is there a platform provided so the swing of the door does not reduce the width of the platform to less than twenty-one inches?
- ❑ Where stairs or stairways exit directly into any area where vehicles may be operated, are adequate barriers and warnings provided to prevent employees stepping into the path of traffic?
- ❑ Do stairway landings have a length measured in the direction of travel that is at least equal to the width of the stairway?

Elevated Surfaces

- ❑ Are signs posted, when appropriate, showing the elevated surface load capacity?
- ❑ Are surfaces elevated more than thirty inches above the floor or ground provided with standard guardrails?
- ❑ Are all elevated surfaces (beneath which people or machinery could be exposed to falling objects) provided with standard four-inch toeboards?
- ❑ Is a permanent means of access and egress provided to elevated storage and work surfaces?
- ❑ Is required headroom provided where necessary?
- ❑ Is material on elevated surfaces piled, stacked, or racked in a manner to prevent it from tipping, falling, collapsing, rolling, or spreading?
- ❑ Are dock boards or bridge plates used when transferring materials between docks and trucks or rail cars?

HAZARD COMMUNICATION

- ❑ Is there a list of hazardous substances used in your workplace?
- ❑ Is there a written hazard communication program dealing with Material Safety Data Sheets (MSDS), labeling, and employee training?

- ❑ Is each container for a hazardous substance (i.e., vats, bottles, storage tanks, etc.) labeled with product identity and a hazard warning (communication of the specific health hazards and physical hazards)?
- ❑ Is there a Material Safety Data Sheet readily available for each hazardous substance used?
- ❑ Is there an employee training program for hazardous substances?

  Does this program include:

  - ❑ An explanation of what an MSDS is and how to use and obtain one?
  - ❑ MSDS contents for each hazardous substance or class of substances?
  - ❑ Explanation of "Right to Know"?
  - ❑ Identification of where an employee can see the employer's written hazard communication program and where hazardous substances are present in work areas?
  - ❑ The physical and health hazards of substances in the work area, and specific protective measures to be used?
  - ❑ Details of the hazard communication program, including how to use the labeling system and MSDSs?

  Are employees trained in the following:

  - ❑ How to recognize tasks that might result in occupational exposure?
  - ❑ How to use work practice and engineering controls and personal protective equipment and how to know their limitations?
  - ❑ How to obtain information on the types, selection, proper use, location, removal, handling, decontamination, and disposal of personal protective equipment?
  - ❑ Who to contact and what to do in an emergency?

As you can see, even if you eliminate questions that obviously do not apply, the number of concerns remaining is still formidable. And while it might be nice to suggest that a trained professional should be responsible for her own safety on the job or in the workplace, the federal government has made it your responsibility as an owner or a manager to see to it that that trained professional has been given the tools and training necessary to ensure that she is able to carry out her responsibility.

TRAINING REQUIREMENTS

As suggested earlier in the text, if employees have a "right to know," then someone else has a responsibility to tell them. That translates as a responsibility to train all employees with regard to the safety hazards they might be exposed to in the workplace and the proper ways to conduct and protect themselves in order to ensure their safety and the integrity of the environment.

Training is an extension of the employee's right to know and should be an integral part of any illness prevention program you consider instituting in order to meet OSHA standards. It should include safety training as well as skills training. Training program guidelines should include, but not be limited to, the following:

1. Determine if training in any specific area is needed.
2. Identify training needs.
3. Identify training goals and objectives.
4. Develop learning activities designed to meet those needs, goals, and objectives.
5. Conduct the training.
6. Evaluate program effectiveness.
7. Redesign the program to improve effectiveness.

This model was designed by OSHA to be equally as effective for a small business with just a few employees as it might be for a larger business with many employees, and despite what you might be thinking at the moment, it is not as difficult to develop or administer as it might look. So let's break it down into the seven elements we've identified and work our way through the requirements of a successful training program

Determining the Need for Training

The first step in developing an effective response to any problem is to determine whether or not the response you have planned is in fact an effective one. In many cases, "training" is a blanket response to any one of a number of problems. However, the increased knowledge, different policies and procedures, and/or deeper understanding that could possibly result from an effective training program may still be ineffective when it comes to eliminating the problem. In some cases, there are other more physical responses, such as the actual removal or containment of a potential hazard, that could provide a more positive result. Albert Einstein suggested that a problem cannot be solved by the same level of awareness that caused it, and creating safety or illness prevention programs to address hazards in the workplace is no exception.

The purpose of training in this regard is not to eliminate all dangers associated with hazards in the workplace, but to ensure that there are no more hazards in the workplace than are absolutely necessary, and that everyone understands what to do in the case of an emergency resulting from one of those hazards. The problems that can be addressed most successfully by training are those that focus on policies, procedures, and work-related rules designed to eliminate or avoid practices that are unsafe and that may result in the possibility of physical harm. In many cases, this can include simply familiarizing employees with the safe practices associated with their work-related tasks, the proper use of the tools and equipment necessary in the performance of those tasks, and/or the incorrect execution of certain work-related tasks.

Identifying Training Needs

If, in the assessment of a problem or potential problem, you determine that the solution can be provided by training, the next step is to determine exactly what kind of training will work best. In order to do that, you have to understand what the "desired result" would be. This is a concept we touched on in *From Intent to Implementation*, another book in this series that is focused on strategic planning. The "desired result" is an imaginary look at your situation and what it would be like if the training you are designing worked perfectly, and then working backwards from there to see exactly what would have to happen in order to make that "desired result" a reality.

Generally, this process begins with an analysis of what the employee is expected to do with regard to a specific task, as well as the best, most efficient, most effective way to get it done. There are procedures for just about everything, and this is no exception. However, the easiest way to approach this task is to monitor the performance of a specific task, documenting all the steps through its successful completion. Once you have established the "best way" to accomplish that task for your facility, you can proceed to find different ways to monitor, manage, and/or eliminate the risks involved by using what you have learned to create a training program designed to help new or even existing employees to meet or exceed your wants, needs, and expectations.

An excellent assessment tool when trying to determine whether or not a specific performance area needs attention can be provided by the employees themselves in response to your questions about work or work-related tasks. A simple question, like "What work-related tasks frighten you?" Or, "During the performance of which tasks are you the most careful or the most anxious?" can tell you a great deal about where hidden dangers may be lurking.

Identifying Training Goals and Objectives

To be most effective, training should focus only on the areas in which the greatest performance improvement can be recognized and achieved. Employees should be made aware of all the steps in a task or a procedure, but training should be reserved for only those tasks or procedures where improvement is required. Time and effort should not be wasted going over those areas of performance where there are no problems or where performance is already adequate or beyond.

Training goals and/or objectives should focus on behaviors: those the employer wants the employee to perform, those the employer wants the employee to perform better, or those the employer wants the employee to stop altogether.

In order to be effective, training results should be measurable, and in order for them to be measurable, there must be a standard applied. That standard does not have to be complicated. In fact, it can be quite simple. If, for instance, safety training revolves around the use of personal protective equipment, it could simply mean that at the culmination of the training "The employee will both know and understand the importance of using the proper eye protection and when such use is appropriate."

This can easily be determined two different ways. The first is by giving the employees a written test. This can serve to reflect the information the employees have been exposed to, while documenting that the training was offered and did take place—as well as what the employees learned and how they may have benefited from that knowledge. The second is simply by observing the employees in the performance of their everyday work responsibilities to see if they are putting what they have learned into practice.

Developing Learning Activities

Learning activities help those being trained demonstrate that they have learned the material you have presented and acquired the skills and abilities necessary to perform their assigned tasks safely and without danger to themselves, anyone else, or the environment.

These learning activities can focus on *how to*; or they can focus on *why to*. *How to* activities generally will be sequential in nature, beginning with the first step required to perform a task regardless of how simple or obvious that task might be. *Why to* activities generally have more to do with philosophy or theory than they do with actual application, but they can be just as important because they can provide the motivation necessary to sit through the *how to*!

In many ways, these learning activities can be classified as passive or active as well. While it is important that the object of this training—the employee, or the student—is engaged at all times, which itself implies a certain level of activity by definition, that activity can take many different forms. Students can watch or listen to a lecture, an audio tape, a video, overheads, or a PowerPoint presentation. Or they can become actively engaged in the actual performance of the tasks they are being trained to perform, either on-line in a virtual environment or at the shop itself.

When you see this icon throughout the book, please take a moment to visit www.trainingbay.com for additional information. The information found there will include training solutions that may take the form of an online companion, sample forms, other manuals, and more.

Activities can be designed and developed for all your employees as a group, or they can be self-paced and focused on the individual. Regardless of how they are designed or delivered, the desired result should be employees who can demonstrate that they have acquired the necessary skills and abilities to perform their job-related responsibilities safely and without incident.

Conducting the Training

The secret to any kind of effective communication is often condensed into the following simple instructions: Tell them what you are going to tell them. Tell them. And then tell them what you just told them. The kind of training we're talking about now is perhaps the best example of how this directive should be applied.

To be effective, you should begin by presenting an overview of what material you will be covering and why. It should be noted that it is critically important for you to provide the student with a pretty good WIIFM—a good explanation of what you are doing and why you are doing it, so the student can walk away with a deep understanding of *What's In It For Me*!

This is best accomplished when the program goals and objectives are explained in terms of their relevance and potential benefit.

Evaluating Effectiveness

The value of any kind of training or education can only be found in its effectiveness: did it accomplish that which it was designed to accomplish?

The easiest way to assess whether or not training is effective is to monitor performance and/or behavior. If the behaviors change in a positive way and performance improves or at least moves in the desired direction, that training could be considered a success. If things remain where they are without any kind of movement, the training program could be classified as ineffective.

There are other ways to measure success in addition to the obvious. One is by surveying employee opinions—although that isn't always the best indicator of effectiveness, as employees may be inclined to tell you only what they think you want to hear. However, listening to what your employees are telling you about the training you have offered or the training they have received, while monitoring their actual performance on the job, can be significantly more revealing. Realistically, the most effective way to evaluate the success of your training efforts would be a combination of surveying employee reactions to the training, measuring employee performance, and monitoring overall performance in the workplace in terms of fewer accidents or improved safety conditions.

Regardless of what you measure or how you measure it, evaluation is critical in order to determine whether or not your efforts were warranted and whether or not you received an adequate return on your training investment.

Improving the Program

Frankly, it really doesn't make sense to do any of this if you aren't truly interested in reducing the risk of injury or accidents, or improving overall working conditions. Compliance alone will not be enough to drive this kind of effort, at least not for very long. You have to be willing to dedicate yourself to safety in the workplace and a hazard-free environment for both your employees and customers—and that means a commitment to continuous improvement.

To continuously improve this kind of process, you will have to revisit it with a critical eye on a regular basis, paying attention to any elements that may have been inadvertently omitted; for areas of performance that are stagnant; and areas that seem to be reverting to older, more comfortable behaviors—or worse yet, deteriorating. You will at least have to *try* to determine if there were elements of the program that were distracting, or unnecessarily confusing, or for some reason unnecessary. And, perhaps most important of all, you will have to make an honest appraisal of what the employees learned or were not able to learn.

In the end, you may have to start the process all over again, beginning with a job analysis to determine whether or not a critical component of the job was somehow left out.

EXAMPLES

Here are a few examples of the kinds of elements that might be included in work-related training of this nature, specifically with regard to personal protective equipment.

Foot and Leg Protection

If you were going to train your employees on the importance of wearing the proper or appropriate protective gear for their feet or legs, here are a few of the program objectives you might want to consider.

First and perhaps foremost, you would want them to leave the training understanding when and why foot and leg protection is important, as well as the hazards that make wearing protective gear prudent, if not absolutely necessary.

You would want your employees to leave the training knowing exactly how specific pieces of the appropriate personal protective gear will protect them; the limits of that protection; proper use of the equipment; how to tell when the equipment may be losing its effectiveness or when it is approaching the end of its service life; and finally, how to take care of it, clean it, and ultimately how to discard it.

Hand and Arm Protection

The same criteria used above can be used here, with only the nature of the equipment itself being different. We would still want to instill in the employees a sense of why the use of this particular kind of personal protective gear is important, as well as the workplace conditions that make using it an effective safeguard.

We would want the employees to understand when the appropriate kind of gloves, gauntlets, or sleeves will protect them and when they won't. We would certainly want them to both know and understand when to wear them, how to wear them properly, and how to determine if fit or function is going to be a problem.

In addition, we would want them to be able to determine when protective equipment of this kind is showing signs of failure or wear, how to clean and care for it, how to disinfect it when indicated, and how to discard it properly when necessary.

As you can see, you can create a kind of template for training development that is uncomplicated and relatively straightforward and that answers all the questions you want your employees to have the answers to.

EMPLOYEE RIGHTS AND THE RIGHT TO KNOW

I should warn you that the following statement is probably not as objective as it could or should be, especially in this context. However, it is based upon experience—my experience—and consequently, will reflect my views on the issue of the right to know.

Most government agencies function with a complete disregard for our system of justice: a system of justice that demands the accused be considered innocent until proven guilty. They lean instead toward a Napoleonic system of justice, whereby the accused is guilty until he can prove himself innocent. OSHA is no exception to this general rule, with a bias that seems to originate from a deep seated belief that given the opportunity, most employers would take advantage of their employees.

While that has not been my personal experience, I can see how and why such a mindset could develop. I'm sure the logic goes something like this: If employers truly cared about the health, safety, and well-being of their employees, the government would not have to demand that employees be made aware of any hazards or dangers associated with the workplace through right to know legislation. The employers would do that willingly, voluntarily, and entirely on their own. But since that does not happen, employers cannot be trusted, and legislation along with government oversight and intervention are required to protect employees from any unnecessary risk.

Realistically, the logic trail I have cited above would probably be difficult to debate, since workers often have not been made aware of hazards associated with the workplace. But I'm not sure the actual reasons for these failures are consistent with those the government would be comfortable citing. My guess is that employers, myself and my family included, probably did not make their employees aware of various hazards in the workplace because those hazards were—for the most part—simply a fact of life: a part of the business, a part of what we did.

You see, I don't think for a moment that we didn't recognize the dangers associated with a career in the automotive service industry! I believe we were very much aware—after all, we certainly knew that a battery could blow up because it generated hydrogen gas (the result of which is illustrated in Figure 2-1), and that battery acid burned.

Figure 2-1 Battery hazards.

We knew that extreme care had to be exercised while working on or around an engine that was running at operating speeds of between 700 and 2,800 revolutions per minute. We understood that sparks can fly and fire can burn, that if you fall you are likely to get hurt, and that explosions can kill or maim or just generally create an unsightly mess.

However, what we weren't aware of were the hidden dangers associated with many of the chemicals and compounds we were using, any more than we were aware of the dangers of leaded fuel, breathing toxic fumes twelve hours a day, or the consequences of releasing Refrigerant

12 into the atmosphere. Nor am I sure that, armed with that knowledge today, we would be as cavalier about that kind of risk as we were then, without knowing.

Shop owners and managers didn't furnish this information to their employees because they weren't aware of the consequences themselves. It was certainly not because they were afraid their employees would leave if someone told them how dangerous working in this environment could be. Shop owners didn't share the information because it wasn't a consideration at all. It was taken for granted—just a part of what we did, period!

If it was a secret at all, it was a secret kept from everyone!

Nevertheless, things have changed. It is the law of the land today, and a right to know is mandated by OSHA standards. That means that we, as responsible business people, must make public the hazards associated with working in an automotive service environment (see Figure 2-2).

Figure 2-2 Employee Right to Know and Material Safety Data Sheet documentation.

That means that as an employee you have the right to a safe and healthful workplace and to:

- Notify your employer or OSHA about hazards in the workplace and the right to request that OSHA keep your name confidential
- Request an OSHA inspection if you believe that there are unsafe and unhealthful conditions in your workplace and for you and/or your representative to participate
- File a complaint within 30 days of discrimination by your employer for making safety and health complaints or for exercising your rights under the OSHA Act
- See any OSHA citations issued to your employer. Employers must post citations at or near the place of the alleged violation
- Expect your employer to correct workplace hazards by the date indicated on the citation and certify that these hazards have been reduced or eliminated

- See copies of your medical records or records of your exposure to toxic and harmful substances or conditions
- Have this notice posted in a prominent place in your workplace
- Be notified if and/or when hazardous materials have been released into the work environment
- Be trained with regard to safety and health hazards related to substances or compounds in the workplace, as well as with regard to the safe and proper handling of these chemicals and compounds, including emergency preparedness and response procedures
- Be trained with regard to reading, understanding, and properly using the content material found in a Material Safety Data Sheet
- Be trained with regard to the specific hazards that exist in the workplace and the nature of those hazards, where the materials are stored, where and how they are used, and what to do in the event they are discharged or released into the workplace environment
- Be trained with regard to the written hazard communication program

Basically, when all is said and done, everyone is responsible for their own personal safety and health, and for the safety and health of those around them—just as everyone is responsible for the protection of the environment. After all, it isn't *your* environment or *my* environment—it's *our* environment.

Right-to-know legislation was enacted to ensure that everyone in the workplace has access to whatever information is necessary and appropriate for any of the potentially hazardous or dangerous chemicals or compounds they are forced to work with or near.

There are three components to the kind of hazardous communication program that is likely to grow out of the right to know: training, labels, and Material Safety Data Sheets. However, the bottom line with regard to any of the chemicals we are likely to come into contact with in our industry is this: if a substance is flammable, corrosive, reactive, or toxic—it's dangerous, and consequently it constitutes a hazard to someone!

All such materials must be properly labeled and handled, packaged, transported, and disposed of. Anything less is a violation of federal law, and in many cases state law as well.

SAFETY PROCEDURES

It may seem patronizing to suggest that there is a *right way* and a *wrong way* to do just about everything, and that being careful and embracing safe practices and procedures as an intrinsic part of your company's culture is nothing more than common sense; but in fact there is a right way and a wrong way to do just about everything, and, unfortunately, there really isn't anything *common* about common sense!

Consequently, bringing safety into the workplace is truly a matter of integrating safe practices into your basic policies and procedures and then ensuring that everyone is properly trained in the safest and most practical way to accomplish every task they will be asked to perform. There is certainly no reason to *assume* that they would just naturally know what that way is—not unless we were the ones responsible for their initial training and acculturation and we told them.

> **ac·cul·tur·a·tion** (ə-kul´chə-rā-shən) *noun* 1. The process by which the culture of a particular society is instilled in a human being from infancy onward. 2. The modification of the culture of a group or an individual as a result of contact with a different culture.

What are these safety procedures?

> **pro·ce·dure** (prə-sē-jər) *noun* 1. A manner of proceeding; a way of performing or effecting something. 2. A series of steps taken to accomplish an end.

Well, there are an endless number of procedures applicable to the hundreds, perhaps even thousands, of tasks we are likely to confront in the course of a normal workday. However, there will always be one procedure that will be the most effective, most efficient, safest, and most appropriate way to successfully accomplish any given task—and that applies to just about anything and everything it is likely we will be asked to do. If we are talking about lifting a vehicle, the lift manufacturer or the Automotive Lift Institute will probably have something to say about what kind of lift should be used, how the vehicle should be positioned on the lift, where the lift points are, and where the pads should be positioned in order to ensure that the vehicle is raised and lowered safely without having it fall off the lift and onto an unsuspecting technician!

If you purchase a new tool or piece of equipment that requires special handling or care, that tool or piece of equipment will generally come with written instructions and warnings to ensure its safe use. But when it comes to handling a hazardous material, there is a lot more to consider than just using that substance properly, safely, and responsibly, and in a manner consistent with the purpose for which it was designed. There are safety issues regarding how and where it should be stored; how and in what it would be best contained; how to safely move it; how to safely and responsibly dispose of it; and what to do if there is a problem with a spill, exposure, or contamination.

We could include a number of these procedures here in the text, but that really isn't what this book is all about. This book is designed to do two things: first, to give you a sense of what safety communication is; and second, to help you understand why it is critically important that you both comply and actively participate. So instead of a litany of safety procedures that may or may not apply to your current reality, let me just share a few examples to help get you started—because that is what a hazardous communication program is really all about: communicating the information you need to develop or adapt procedures that will control and/or manage these hazards in a safe, prudent, and efficient manner.

1. Clean up any spills immediately after they occur in a manner appropriate for the kind of liquid that has been spilled—exercising good judgment and in accordance with all posted shop regulations and in conformance with state and federal law.
2. Follow appropriate safety procedures while working on a vehicle that is running, especially if there is no one at the controls—use wheel restraints when appropriate and ensure that the emergency brake is applied and will restrain the vehicle from moving. Ensure proper ventilation.
3. Use personal protective equipment for any job you are doing in which there is a possibility of injury, including safety glasses or goggles, gloves and/or gauntlets, face protection or a hard hat, safety shoes or boots, and/or leggings.
4. Never use power tools, or any tools for that matter, if the safety guards have been removed or purposely disabled, etc.

At the beginning of this passage I suggested there was nothing common about common sense. Well in my opinion, there isn't. However, if you take a moment to think about it—a good safety communication program may just help make common sense a little bit more common.

MEETINGS

If a safe and hazard-free workplace is our ultimate objective, effective training delivered in conjunction with clear and open communication will almost certainly prove to be the most efficient way to get there. And the best way to deliver that kind of training and facilitate that kind of communication is at regularly scheduled team or staff meetings, where safety- or hazard-related topics have been placed on the agenda, or at meetings specifically designed to cover nothing other than safety-related issues.

Depending upon the size of your business and/or the number of employees, periodic safety meetings may be mandated by law. If they are, the purpose of the meetings should be focused on safety, safety training, and answering any questions related to hazardous materials handling, storage, and/or disposal, or any questions related to safety in the workplace.

There are an endless number of subjects appropriate to such meetings, including the content of Material Safety Data Sheets, hazards specific to the kind of work your company may be engaged in, safety procedures, the care and use of new equipment, and any new safety practices or procedures.

Everyone should be aware of the fact that these meetings have a specific purpose: to communicate information that will ultimately result in a safer and more hazard-free environment and that in order for both the business and the program to be in compliance with OSHA standards, the results must be measurable. That means the people you have trained must demonstrate their new proficiencies, their new skills and increased knowledge, through practice and theory; and then in turn these results must be documented.

The meetings should have a formal agenda, occur on a regular basis (once a month as dictated by law, if applicable), and culminate in a written and/or practical test. As a facilitator, you should do everything you can to encourage everyone to communicate any new hazards that may compromise safety in the workplace or any other safety-related concerns. This would also be the appropriate time to review the results of any safety or facility inspections that may have taken place recently, the relevance of any previously delivered safety training, or suggestions for future training needs or hazard management, abatement, or control.

For a more in-depth discussion of meetings in general, see Chapter 3 in *Building a Team*, another book in the *Automotive Service Management* series.

SUMMARY

You can find the root of the word *communicate* in the Latin *communis*, or *common*: as in achieving a common understanding. Consequently, you could say that the ability to effectively *communicate* begins with a sense of exactly what information needs to be shared to facilitate this common understanding throughout your organization.

Perhaps the best place to start is with a comprehensive assessment of the workplace in order to determine what hazardous material/safety communication issues will require your immediate attention.

Although not specific to our industry, the general self-assessment worksheet included in this chapter is one of the most comprehensive you will ever find. Completing it will identify those areas in which education, information, and training will be necessary to achieve OSHA's primary objective: a safe and hazard-free work environment.

Training, in order to be relevant, must be appropriate. That means you need to both know and understand what information must be shared: what your employees don't know they don't know (Stage 1). Once you have identified what they don't know, you will know what you have to share with them (Stage 2). The third step in developing this kind of training is either finding commercial solutions available in the marketplace or creating a set of *learning activities* that have as their *desired result* a safe and hazard-free workplace in which everyone in the organization knows what they need to know in order to do what they need to do!

The previous two steps in the training process should deal with actually delivering the training and then evaluating its effectiveness: whether or not the training meets the program's goals and objectives. These goals and objectives should relate to the specific rights enumerated earlier, the rights to:

- Notify your employer or OSHA about hazards in the workplace
- Confidentiality
- An OSHA inspection if you believe that there are unsafe and unhealthful conditions in your workplace, and the right for you and/or your representative to participate
- File a complaint within thirty days of discrimination by your employer for making safety and health complaints
- See OSHA citations issued to your employer

- Have citations posted at or near the place of the alleged violation
- Have your employer correct workplace hazards in a responsible and timely manner
- Have access to copies of your medical records or records of your exposure to toxic and harmful substances or conditions
- Be notified if and/or when hazardous materials have been released into the work environment
- Be trained with regard to safety and health hazards related to substances or compounds in the workplace as well as the safe and proper handling of these chemicals and compounds, including emergency preparedness and response procedures
- Be trained to read, understand, and properly use Material Safety Data Sheets
- Be trained with regard to the specific hazards that exist in the workplace and the nature of those hazards, where the materials are stored, where and how they are used, and what to do in the event they are discharged into the workplace
- Be trained with regard to the written hazard communication program

As impossible as this may seem at first, this training should also include your shop's policies and procedures regarding *safety* and *safe practices* within the workplace—in other words, the *right* way, the *safe* way, to perform each of the countless tasks we are asked to perform every day.

Realistically, the best way to accommodate this kind of training is within the context of regularly scheduled weekly or monthly shop meetings. These meetings should include everyone on staff—certainly, everyone with a need to know. And they should incorporate all of the key points with regard to successfully organizing and orchestrating meetings discussed in Chapter 3 of the fifth book in this series, *Building A Team*.

The most important thing for all of us to remember about any of these management responsibilities is that they are ours and ours alone, regardless of how we might feel about them. Consequently, discharging these responsibilities should be our only concern.

What management responsibilities will you consider your first priority? How will you go about attacking them? Will you institute a series of safety communication-related meetings? If you do, what are some of the topics you might consider for each meeting? Take a moment to write down your answers in the space provided.

Thoughts

Actions

Results

CHAPTER 3

Training Requirements

INTRODUCTION

This chapter will explore training requirements, with specific attention paid to the regulatory requirements involved in training, such as:

- General criteria
- Core criteria
- Hearing conservation and safety
- The environment
- Emergencies
- Workplace violence

REQUIRED REQUIREMENTS

Reading federal guidelines can be about as exciting as watching paint dry or grass grow. However, that is sometimes the only way to access the material you need, unless someone like me does it for you.

To give you a sense of just how comprehensive, exhausting, and exasperating these requirements can be, we're going to draw material from the *Training Requirements in OSHA Standards and Training Guidelines*, OSHA 2254, 1998 (revised).

I will not be providing you with a word-by-word commentary on the entire text in the interest of time, space, and your mental well-being. However, I will try to give what I believe you will need to know in order to be both successful with your program and in compliance, along with a loose translation as we try to make it as understandable as we can. Most government documents follow a basic template that begins with the identification and authentication of what you are looking at.

In this case, the header tells you that the material was copied directly from the OSHA web page for 29 CFR 1910.1200, January 12, 1998, related to training guidelines that are suggested but not mandatory, to be used as a template or model to help you develop a program of your own. It also reinforces the fact that because these are generic guidelines, training more specific to the location and nature of the materials being handled and the hazards your employees will be exposed to *will* or *may* become necessary. If you look carefully you will see that it conforms to the same basic formula that we have been following throughout the series:

Tell them what you are going to tell them. Tell them. Tell them what you just told them.

The following is a description of the fundamental requirement being addressed, which in this case is hazardous materials training and the delivery of pertinent information:

> 1910.1200—Hazard Communication
> - Standard Number: 1910.1200
> - Standard Title: Hazard Communication
> - SubPart Number: Z
> - SubPart Title: Toxic and Hazardous Substances
>
> (h)
> "Employee information and training."
>
> (h)(1)
> Employers shall provide employees with effective information and training on hazardous chemicals in their work area at the time of their initial assignment, and whenever a new physical or health hazard the employees have not previously been trained about is introduced into their work area. Information and training may be designed to cover categories of hazards (e.g., flammability, carcinogenicity) or specific chemicals. Chemical-specific information must always be available through labels and material safety data sheets.

SUGGESTED GENERAL CRITERIA

Telling someone what you are going to tell them, and sometimes even telling them, isn't really good enough if you want to ensure that they really understand what it was you just told them. In many cases, it is critical that the special terms you are using or the regular words you are using in a special way are clearly defined.

You may have noticed that in this particular book and throughout this series there are a number of call-outs with definitions in them. They are there to ensure that the word I'm using is understood in the context of how it is being used. As a part of the "General Criteria" here, I have included a group of definitions borrowed from another section in the "Training Requirements in OSHA Standards and Training Guidelines" as an example.

> **Definitions**
>
> "Competent" means possessing the skills, knowledge, experience, and judgment to perform assigned tasks or activities satisfactorily as determined by the employer.
>
> "Demonstration" means the showing by actual use of equipment or procedures.
>
> "Hands-on training" means training in a simulated work environment that permits each student to have experience performing tasks, making decisions, or using equipment appropriate to the job assignment for which the training is being conducted.
>
> "Initial training" means training required prior to beginning work.
>
> "Lecture" means an interactive discourse with a class led by an instructor.
>
> "Proficient" means meeting a stated level of achievement.
>
> "Site-specific" means individual training directed to the operations of a specific job site.
>
> "Training hours" means the number of hours devoted to lectures, learning activities, small group work sessions, demonstrations, evaluations, or hands-on experience.

As you can see from the words that have been chosen and the definitions provided, these terms really are critical to clear and comprehensive understanding.

Following this general information, the material will usually get far more specific.

SUGGESTED CORE CRITERIA

This is where we get specific with regard to exactly what information your employees will be informed of. If this information were provided in everyday English instead of governmental legal-speak, it would be broken down and presented something like this:

The Requirements ... in English

If you were going to translate the passage I transcribed a few headings above into more readable, more understandable English, it would probably look something like this:

> The requirements of this section pertain to any operations in the workplace where hazardous chemicals are present, as well as the location and accessibility of the hazard communication program including the required lists of hazardous chemicals and the material safety data sheets required by this section.
>
> Realistically, any chemical that is flammable, corrosive, reactive, or toxic should be considered hazardous or potentially hazardous.

With Regard to Training

Training should include, at the very minimum, what hazardous materials are, what hazardous wastes are, how a hazardous material becomes a hazardous waste, and ways to detect the presence or release of these hazardous materials in the workplace, such as visual inspection, continuous monitoring devices (used in facilities that store and dispense motor vehicle fuels), and/or by fumes or odor.

With Regard to Chemical Related Health Hazards

Training will also include the different ways an employee can manage to protect himself from these hazards, including the specific policies and procedures introduced by management to protect all employees from exposure to dangerous chemicals. These include, but are not limited to, safe and sound work practices, emergency procedures, and the appropriate personal protective equipment to be used, as well as the details of the hazard communication program developed by the employer, an explanation of the labeling system, and the material safety data sheet. Finally, the training will also identify where and how the employee can obtain and then utilize this information.

But instead of clear English, the material is presented like this:

> (h)(2)
> "Information." Employees shall be informed of:
>
> (h)(2)(i)
> The requirements of this section;
>
> (h)(2)(ii)
> Any operations in their work area where hazardous chemicals are present; and,
>
> (h)(2)(iii)
> The location and availability of the written hazard communication program, including the required list(s) of hazardous chemicals, and material safety data sheets required by this section.
>
> (h)(3)
> "Training." Employee training shall include at least:
>
> (h)(3)(i)
> Methods and observations that may be used to detect the presence or release of a hazardous chemical in the work area (such as monitoring conducted by the employer, continuous monitoring devices, visual appearance or odor of hazardous chemicals when being released, etc.);
>
> (h)(3)(ii)
> The physical and health hazards of the chemicals in the work area;
>
> ..1910.1200(h)(3)(iii)
>
> (h)(3)(iii)
> The measures employees can take to protect themselves from these hazards, including specific procedures the employer has implemented to protect employees from exposure to hazardous chemicals, such as appropriate work practices, emergency procedures, and personal protective equipment to be used; and,
>
> (h)(3)(iv)
> The details of the hazard communication program developed by the employer, including an explanation of the labeling system and the material safety data sheet, and how employees can obtain and use the appropriate hazard information.

Which do you think would be easier to absorb, assimilate, and/or put into practice? Nevertheless, the training requirements then proceed with the following guidelines:

Employee Information and Training

Once again, the information begins with a header whose purpose it is to tell you what you are about to be told. In this case, we know it will be focused on *"Employee Information and Training"*, especially with regard to exposure to hazardous chemicals.

> Advisory Guidelines for Employer Compliance
>
> Standard Number: 1910.1200 App E
> Standard Title: Guidelines for Employer Compliance (Advisory)
> SubPart Number: Z
> SubPart Title: Toxic and Hazardous Substances
>
> C. "Employee Information and Training"
>
> Each employee who may be "exposed" to hazardous chemicals when working must be provided information and trained prior to initial assignment to work with a hazardous chemical, and whenever the hazard changes.

We are fortunate, because in this case the authors have defined exactly what they consider exposure to be.

> "Exposure" or "exposed" under the rule means that "an employee is subjected to a hazardous chemical in the course of employment through any route of entry (inhalation, ingestion, skin contact or absorption, etc.) and includes potential (e.g., accidental or possible) exposure." See paragraph (h) of this section for specific requirements.

It then goes on to identify how this kind of training can be categorized—by *individual chemical* or by *categories of hazards*, and the importance of education and training in this critical area. In both cases, the MSDS can prove a significant resource just as proper labeling can.

The key objective here would be to have everyone in the organization able to understand what they are reading on the MSDS or what exactly the label is trying to tell us.

> Information and training may be done either by individual chemical, or by categories of hazards (such as flammability or carcinogenicity). If there are only a few chemicals in the workplace, then you may want to discuss each one individually. Where there are large numbers of chemicals, or the chemicals change frequently, you will probably want to train generally based on the hazard categories (e.g., flammable liquids, corrosive materials, carcinogens). Employees will have access to the substance-specific information on the labels and MSDSs.

It may seem odd, translating English into English, but when dealing with legal-speak it will almost always be required—especially if a significant number of those trying to understand and comply with the standards and regulations may not be comfortable with English as their first language.

Take for instance, the following paragraph. It reinforces the importance of information availability, but then recognizes that just having the information available isn't enough. In order to use that information, in order to make it relevant, "effective training" is required to teach the employee not only what is being communicated, but where the information can be found, what it means, and how to access and utilize it. That means actually developing an appreciation for the hazards or potential hazards to be found in the workplace, and a plan to ensure the safe handling, use, management, and disposal of any dangerous or hazardous substances.

> Information and training is a critical part of the hazard communication program. Information regarding hazards and protective measures are provided to workers through written labels and material safety data sheets. However, through effective information and training, workers will learn to read and understand such information, determine how it can be obtained and used in their own workplaces, and understand the risks of exposure to the chemicals in their workplaces as well as the ways to protect themselves.

The suggestion here is that ignorance is the problem and that the right information, delivered and explained intelligently in an environment in which the student is comfortable and secure—which is a pretty good explanation of what training is all about anyway—is the solution.

> A properly conducted training program will ensure comprehension and understanding. It is not sufficient to either just read material to the workers, or simply hand them material to read. You want to create a climate where workers feel free to ask questions. This will help you to ensure that the information is understood.

By restating the purpose of our efforts, we are reminded that our goal is a safe and hazard-free environment and that if successful, the policies, procedures, information, and training that grow out of our program will result in an informed and responsible workforce and a workplace that is both safe and secure.

> You must always remember that the underlying purpose of the HCS is to reduce the incidence of chemical source illnesses and injuries. This will be accomplished by modifying behavior through the provision of hazard information and information about protective measures. If your program works, you and your workers will better understand the chemical hazards within the workplace. The procedures you establish regarding, for example, purchasing, storage, and handling of these chemicals will improve, and thereby reduce the risks posed to employees exposed to the chemical hazards involved. Furthermore, your workers' comprehension will also be increased, and proper work practices will be followed in your workplace.

With Regard to Training Internally

In all honesty, this next section is fairly straightforward—something I never thought I would be able to say about anything published by the government—and that is a good thing, because it is an important section detailing exactly what will be expected of us should we choose to build an internal training program ourselves. But perhaps even more importantly, it goes on to tell you where you can get the support materials you may want or need to build such a program yourself. That is, *if* you are able to translate it!

> If you are going to do the training yourself, you will have to understand the material and be prepared to motivate the workers to learn. This is not always an easy task, but the benefits are worth the effort. More information regarding appropriate training can be found in OSHA Publication No. 2254, which contains voluntary training guidelines prepared by OSHA's Training Institute. A copy of this document is available from OSHA's Publications Office at (202) 219-4667. In reviewing your written program with regard to information and training, the following items need to be considered:

1. Designation of person(s) responsible for conducting training; the "compliance officer";
2. Format of the program to be used (audiovisuals, classroom instruction, etc.);
3. Elements of the training program (should be consistent with the elements in paragraph (h) of this section); and,
4. Procedure to train new employees at the time of their initial assignment to work with a hazardous chemical, and to train employees when a new hazard is introduced into the workplace.

The written program should provide enough details about the employer's plans in this area to assess whether or not a good faith effort is being made to train employees. OSHA does not expect that every worker will be able to recite all of the information about each chemical in the workplace. In general, the most important aspects of training under the HCS are to ensure that employees are aware that they are exposed to hazardous chemicals, that they know how to read and use labels and material safety data sheets, and that, as a consequence of learning this information, they are following the appropriate protective measures established by the employer. OSHA compliance officers will be talking to employees to determine if they have received training, if they know they are exposed to hazardous chemicals, and if they know where to obtain substance-specific information on labels and MSDSs.

There are five critical issues covered in the material that appears above and reinforcing them won't hurt our chances of succeeding. They are:

- The responsibility to assign ownership for the program or any part of the program
- Identifying what the training will consist of (elements)
- Establishing how the training will be delivered and the media (format) to be used
- The appropriate training procedures for bringing a new-hire up to speed with regard to safety in the workplace, and/or the appropriate training procedures when a new hazard is introduced into the workplace.

In addition to all that, however, the material implies that a good deal of this training program must exist in writing as well as in principle. This written program will be evaluated for both its content and effectiveness, if for some reason you find yourself sitting across the table from an OSHA investigator. That means that anyone exposed to your training program will be expected at the very minimum to understand that:

- They are or have been exposed to hazards in the workplace.
- What those hazards are.
- How to recognize and identify them.
- What role the material safety data sheets and labels play.
- How to read them.
- And how, as a consequence of the training, they were able to apply the appropriate behaviors to protect themselves.

Whether precipitated by a conference or a confrontation regarding a safety issue or hazard in the workplace, you want to be able to answer any OSHA investigator's questions in a positive way by being able to document all of the above.

> The rule does not require employers to maintain records of employee training, but many employers choose to do so. This may help you monitor your own program to ensure that all employees are appropriately trained. If you already have a training program, you may simply have to supplement it with whatever additional information is required under the HCS. For example, construction employers that are already in compliance with the construction training standard (29 CFR 1926.21) will have little extra training to do.

As suggested in other books in this series, if you don't have a written record that something happened, it didn't happen, so document everything. And I would suggest that once documented, you should keep those records forever if you can, but for a minimum of seven years.

> An employer can provide employees information and training through whatever means are found appropriate and productive. Although there would always have to be some training on-site (such as informing employees of the location and availability of the written program and MSDSs), employee training may be satisfied in part by general training about the requirements of the HCS and about chemical hazards on the job which is provided by, for example, trade associations, unions, colleges, and professional schools. In addition, the previous training, education, and experience of a worker may relieve the employer of some of the burdens of informing and training that worker. Regardless of the method relied upon, however, the employer is always ultimately responsible for ensuring that employees are adequately trained. If the compliance officer finds that the training is deficient, the employer will be cited for the deficiency, regardless of who actually provided the training on behalf of the employer.

And finally, the last section addresses more principles. The first is the fact that you can purchase any kind of information or training that is available for the purposes discussed above. You can purchase videos, send your employees to school, subscribe to a safety newsletter, or access training provided by your trade association or business group. It doesn't matter, as long as the program requirements have been fulfilled. And the second and perhaps more important principle is that the ultimate responsibility for everything we have been talking about can be found at the feet of the business owner.

The Sin of Omission

With everything else that was covered in the previous sections, it seems almost ludicrous to talk about what wasn't there, but that is exactly what we are about to do.

One of the most positive aspects of writing these books has turned out to be the review process designed to ensure that they are accurate, relevant, and meaningful. In reviewing this book, a question was raised about bloodborne pathogens and whether shop policies and procedures and internal training programs, especially training programs relating to the administration of first aid, should address this critical issue. The question is a valid one, and the answer is, let paranoia dictate your response—recognizing that the absence of the appropriate training programs, polices, and/or procedures would be harder to explain than whether or not the ones you had established were adequate.

In this case, your actions would at least suggest that you were trying to be consistent with the intent of the standards and statutes.

Lacerations and bleeding are almost unavoidable in our industry. And although I'm not aware of any specific legislation relating specifically to this problem and the automotive service industry, why take the chance of providing the legal profession with a test case? Policies, procedures, and the appropriate first aid training should be covered within your Safety Communication Program, since the contraction of a potentially lethal bloodborne disease at work could prove devastating.

For more information, please refer to the listing in the "References and Resources" section at the end of this guide.

HEARING CONSERVATION AND SAFETY

In a way, writing this book has been more than just a little frustrating, especially this particular section. You see, I came into this industry at a time when no one even thought about the subjects that have so far filled the pages of this book. I didn't wear safety glasses to protect my eyes from the dirt or debris associated with what we do because no one I was ever exposed to, no one I ever watched, no one who might have ever been a role model wore them. I didn't wear gloves to protect my skin from contacting potentially dangerous chemicals because no one else wore them and because, frankly, they were so primitive and so user *un*-friendly, you could not wear them and still *feel* what you were doing. And I didn't wear the kind of ear protection illustrated in Figure 3-1 for all the same reasons.

Figure 3-1 Protecting your hearing.

Today, I have to have someone else road test some of the vehicles brought to the shop for noise complaints, because I already know that if it is a noise within a certain high frequency range, I'm just not going to be able to hear it. I have also found myself unable to go for an MRI because of the possibility that there might still be some tiny metal particles inside my eyes or lurking someplace else in my body because of the work I did, the risks I took, and the dangers I confronted.

It's not like we didn't know or weren't aware. We all knew that we were working in a dangerous environment, and we had a pretty good idea of the dangers we were facing and the risks involved. Or at least we thought we did. Today, we have more information at our disposal than we have ever had before. We can't hide from it. In fact, we should embrace it, hoping to spare those who follow us the same risks we faced.

The noise level in most traditional automotive repair facilities is literally deafening! Engines running with and without exhaust, air tools blasting away at full tilt, grinders grinding and compressors going all day—even the hidden, almost inaudible noise of the electric ballasts used for fluorescent lighting is enough to warrant a sigh of relief when you finally close the lights as you leave at night and realize just how loud something you don't hear can be!

There are a number of different ways you can protect your hearing, just as there are dozens of different ways to address all the other hazards we face at work. There are countless ways both to conserve your hearing and to protect your ears from the noise levels at the shop, beginning with insulating the building and/or providing ear protection.

Our shop area is nothing more or less than a big, rectangular metal box. We work in southern California, where the summers can get unbearably hot. For twenty years we suffered through the heat because we were told there was no alternative and that there was nothing we could do. Noise reverberated around inside that box at a horrific level, but like everything else, we got used to it eventually. Then, a few years ago, we found a way to insulate the building by hanging the insulation just below the ceiling. The first thing we noticed when the workers were finished was not the change in temperature. The first thing we noticed was the difference in the noise level! It was quiet. Or at least it was *quieter*, and everyone in the place noticed it.

It isn't enough just to provide ear protection or to passively suggest that your employees wear it. It isn't enough to try to control the noise levels by insulating with sound-deadening materials and then ignoring the potential danger because you did all you feel you could do. It is your responsibility as the owner or manager of your facility to do at least that and more. You really should focus your energy and your efforts on providing your employees the information they need to recognize the risk high noise levels may pose and by educating them about the different ways in which they can protect themselves.

ENVIRONMENTAL CONCERNS

If you have been working your way through this series from the beginning you already have already figured out that I'm a big fan of a good *Why?*—the most important reason fueling any action, particularly something difficult or arduous. If no one can convince you that where you are isn't really where you need to be, there isn't any reason to change anything. It takes a good *Why* to suffer through all the effort, anxiety, and discomfort that accompany even the most minor changes.

For almost the entirety of the Industrial Age humans have lost sight of the fact that we do not have dominion over all we survey, that we are merely custodians of this little rock of ours. However, as we have seen changes in our environment, in the quality of the air we breathe, the water we drink, the foods we eat, the oceans we depend upon for our very existence, even in the climactic conditions we face around the globe, we are starting to realize just how important our contributions to this fragile ecosystem we call our environment can be.

An argument can be made that we are the first wave of assault troops in the battle to protect the environment, the last line of defense against those who add to the problem by neglecting and abusing their vehicles. If we are to be good soldiers, we need to understand what we can do, at least insofar as the service, maintenance, and repair of the automobiles and trucks we work on is concerned, to do our part—our fair share.

Realistically, there are not a lot of provisions that demand that we recognize the implications of our actions from an environmental perspective. In fact, most of the regulations are punitive and have more to do with abuse than conservation: what happens after you let anti-freeze/glycol drain into the ground water, discharge Refrigerant 12 into the atmosphere, or pour waste oil down the drain. Nonetheless, it is your responsibility to see that everyone in your organization is complying with governmental regulations, especially with regard to the environment and environmental concerns.

For more information on what compliance actually entails and resources available to help you meet those requirements, please look at the listing in the "References and Resources" section at the end of this guide.

EMERGENCY PLANNING

Accidents happen. Regardless of how much we know or how hard we try to avoid danger or a potential hazard, accidents just happen. That's why they call them accidents, I suppose. The question before us is, what do you and I do when accidents happen? How do we respond?

You see, there may be some performance areas of your life in which you are seduced into offering a vague, almost non-responsive answer. But this isn't going to be one of them. The government, specifically the Occupational Safety and Health Administration, has seen to that. So this will not be an if question: a question you answer *if* you so choose. This will be a *how, what, why, where, when, what,* answer—an answer that, because of who asked the question, demands a thoughtful and precise answer. And since it was the government that asked the question, nothing less will do.

Since we have already established that accidents *will* happen, and that many times these accidents will spiral into an emergency, all that is left to do is define what constitutes an emergency, and then discuss realistic ways in which we might respond.

The United States Department of Labor suggests (*www.osha.gov/pls/oshaweb/owadisp.show_document?p_table=STANDARDS&p_id=10114&p_text_version=FALSE*) that the emergency plan should address emergencies the owner might realistically expect. Consequently, a California automotive shop owner or manager should have a plan that takes into account the fact that every once in a while the ground moves—we have earthquakes—and when it does, just about everything that isn't secured in some way is likely to move with it. A garage owner in Florida really ought to have a plan that encompasses hurricanes or severe tropical storms, and one in Montana or North Dakota should probably dedicate a few minutes for some serious thought regarding blizzards. In addition, it would be safe to say that some of us really should be thinking about floods, while all of us have to worry about fires. Realistically, we all have to worry about toxic chemicals and what happens if they are released into the atmosphere or spill out on to the ground.

In the great scheme of things, what kind of emergency it is really doesn't matter—all that matters is that you can prove that you have taken the time and made the effort to think about the most effective and prudent ways you can protect the health, safety, and welfare of those you are responsible for—particularly your employees—and that you have done something about it: namely, established a plan that reflects the situation above and then trained everyone involved with regard to what they are supposed to do.

For those employers with ten or more employees, that plan "must exist in writing, be kept in the workplace and be available to employees for review." For those with less than ten, the plan can be communicated orally, but frankly, I would still take the time to put it in writing, just to be able to prove that a plan really did exist. It isn't that difficult to do when you realize that whether communicated orally or by writing, it must have the same basic procedures designed to accommodate the following:

- Reporting a fire or other emergency (You don't have to do that with an earthquake because it is unlikely that one large enough to constitute an emergency will go unnoticed by anyone, no matter how oblivious they might be!)
- Evacuating the workplace in a safe and orderly fashion, including approved exit routes and their locations
- By any employees who must stay behind for any reason, and for those responsible to ensure their safety
- Roll-call employees to ensure everyone is accounted for after an evacuation
- Employees responsible for rescue or medical duties
- The name of everyone responsible to provide this information about the plan or what they are supposed to do in the event of an emergency

Additionally, the plan must provide for a way to alert employees to the emergency, training with regard to *who* is supposed to do *what* to *whom, when* and *how*, and the active and on-going review of the plan to ensure that everyone really does know and remember what they are supposed to do if there is an emergency.

FIRE PREVENTION PLAN

Any of these examples may be one of those rare times someone comes up to you, the owner of an automotive service business, and says, "I'm from the government, and I'm here to help you!"—and actually means it.

Regardless of how careful you are, or think you are, fire is and always has been a real danger in our environment. So having a plan in place to address this very real hazard is nothing if not prudent. And once again, the same rules regarding oral and written plans and number of employees apply. However, once again, I will suggest that regardless of how big or small you are, you have a written plan in place along with the training that goes with it. The worst thing that may happen is lower insurance rates!

Now, here are the basic elements that should be included in such a plan:

- List of all major fire hazards
- Proper storage and handling procedures for all hazardous materials and wastes

- Ignition sources (or possible ignition sources) along with their controls
- The fire protection equipment necessary for each type of threat as illustrated in Figure 3-2
- Procedures to monitor and manage accumulated flammable and combustible waste
- Procedures to prevent the accidental ignition of combustible materials
- The person(s) responsible for maintaining emergency fire protection equipment
- The person(s) responsible for monitoring, managing, and controlling fuel source hazards
- The person(s) responsible for educating and training everyone else

Figure 3-2 Fire prevention equipment.

Remember, it is your responsibility as an owner or manager to identify potential fire hazards for anyone entering your employment, as well as to provide a review of the established fire protection plan along with the new hire's role and responsibilities with regard to fire safety.

PREVENTING WORKPLACE VIOLENCE

I'm not sure workplace violence should be included along with all the other hazards in a comprehensive hazard communication program. I'm not sure it is the same kind of hazard that even some of the most dangerous hazards might be. Not because it's any less dangerous. That's not it at all! I'm just not clear on whether being confronted by a disgruntled or frustrated employee or former employee, customer or former customer, vendor or former vendor is the same thing as breathing noxious gases, toxic fumes, asbestos particles, biting your finger nails after they've been covered in grease, or washing with solvents we've used over the years.

Please don't misunderstand. I'm not minimizing either kind of hazard. I just think they are very different in a number of ways and for a number of reasons.

Asbestos was included in friction materials for years because, frankly, there really isn't anything better for the job than that natural fiber. Unfortunately, under a number of different circumstances, it has proven itself severely carcinogenic. We were exposed to it for years without realizing the dangers involved—and really, that is a significant part of why there is a Right to Know program and a Hazard Communication Standard in the first place. But asbestos doesn't have a choice about being a hazard, it just is.

We were exposed to lead in gasoline for almost all the early years I was in the industry because of its ability to control pre-ignition and spark detonation. But like asbestos, lead doesn't have a choice about being poisonous, it just is. In both cases, as soon as the health risks were identified, documented, and confirmed, the products were monitored, removed, or controlled almost immediately.

In virtually every other instance I can think of that involves a hazardous substance, the initial exposure was involuntary: a fact of life, because we needed the substance in order to do what we do. And in every instance I can think of, the substance did not have a choice. It wasn't lethal because it wanted to be, it was lethal because of its chemical composition.

In a way, it is rather sad that workplace violence is included in the same category as all the other dangers we have been talking about thus far, because people *do* have a choice. People don't *have* to resort to violence. And things never *have* to go that far. However, having said that, having made the distinction between the human element and a host of chemical elements, workplace violence has become a fact of life for almost all of us. Even if we have so far managed to avoid it personally, we can't escape it, because it is all around us. We drown in it every morning, afternoon, and evening as the media pours a constant stream of such violence over us in our national and local newspapers and magazines, on television and on the radio, and as the subject matter of the books we read and the movies we see.

As a result of the violence brought to the workplace by distraught, frustrated, angry, or disturbed individuals, it has become necessary to outline policies and procedures to follow, as well as education and training to help minimize the possibility of being confronted by such behavior on the job. Forget about your employees' *right to know*, which makes it your responsibility to remind them that someone *could* bring violence to the workplace. Knowing what to do if or when that occurs has become a personal responsibility that all of us must accept in light of what we have witnessed over the past few years, especially with regard to violence in the workplace.

Where do you do go? What do you do? Who do you call? Do you confront someone who is enraged or irrational? Or do you exit the building, alerting everyone you can as you leave? If someone has a weapon, do you try to take it away? Or do you retreat so that you can call the police? Do you meet force with force, violence with violence? Or do you accept the reality of our time and plan for a possibility we all hope can be avoided?

Like so many of the other hazards we have been discussing, the key to avoiding a dangerous or life-threatening situation is information, awareness, and planning—information about what workplace violence is all about, and the early signs of danger that could help you help someone working for you or with you *before* there is a crisis. Once informed, you have—everyone in your organization has—a responsibility to remain aware of your environment for those early warning signs. And you need to establish a plan, complete with the appropriate policies and procedures, designed to deal with the possibility of such violence as quickly and efficiently as possible, while minimizing the risk.

But is physical violence the only kind of violence likely to rear its ugly head in the workplace, *your* workplace?

Other Kinds of Violence

We have a responsibility to ourselves and those around us to recognize that pain can be inflicted and harm can be suffered without the use of an actual weapon. Harassment, both verbal and sexual, can and often does find its way into the workplace if we are not careful. People can be verbally abusive or physically confrontational without ever actually touching the other person, and the results can be just as harmful.

No one should have to go to work only to suffer the intimidation of a bully. No one should have to spend any of their work day afraid or uncomfortable, especially if their discomfort occurs at the hands of another. It is our ultimate responsibility as the owner or manager of the business to be aware at all times of this kind of violence in the workplace—and if or when it becomes apparent that it is present, to act upon it immediately.

Planning and Execution

The key here is to recognize that workplace violence can take many forms and that regardless of the nature of that kind of violence or abuse we have a responsibility to remain aware of our environment in order to ensure that it is safe. We have to create an environment in which there is no room for such a hazard—and that will take information, education, and more than a little understanding. Because whether you believe this is your responsibility or not, it is.

SUMMARY

At the very heart of any document that has the word *communication* as part of its title should be the ability to *convey information* or to *reveal clearly* the information you are trying so desperately to share.

It is this ability to exchange ideas, this ability to express new concepts clearly enough for others to understand, that should be the goal of any successful training program.

com·mu·ni·cate (kə-myoo-ni-kāt) *verb* com·mu·ni·cat·ed, com·mu·ni·cat·ing, com·mu·ni·cates *verb, transitive* **1. a.** To convey information about; make known; impart. **b.** To reveal clearly; manifest. *verb, intransitive* **1.** To have an interchange, as of ideas. **2.** To express oneself in such a way that one is readily and clearly understood by all.

Within the context of safety communication, the training requirements are fairly specific because the information, ideas, and concepts you are compelled to share with all of your employees are clearly defined and consequently fairly precise. They should include: what hazardous materials are, what hazardous wastes are, how a hazardous material becomes a hazardous waste, and ways to detect the presence or release of these hazardous materials in the workplace—such as visual inspection, continuous monitoring devices (used in facilities that store and dispense motor vehicle fuels), and/or by fumes or odor.

The training should also include: different ways employees can protect themselves from hazards, safe work practices, emergency procedures, appropriate personal protective equipment, details of the hazard communication program, an explanation of the labeling system, material safety data sheets, and how and where employees can obtain and utilize all this information.

The basic assumption of an effective training program is that ignorance is the problem and that the right information, delivered and explained intelligently in a comfortable and non-threatening environment, is the solution.

If you are going to provide safety communication training internally, there are five critical areas you must consider: someone must be ultimately responsible for the program, the specific elements of the training must be identified, a delivery mechanism must be selected, the appropriate media must be chosen, and the most effective means of bringing a new employee up to speed with regard to safety in the workplace must be identified—and when a new hazard is introduced into the workplace, the most effective means of educating everyone on staff must be identified as well.

In addition to all this, your program must be documented. In other words, it must exist in the physical universe with a copy available for evaluation if and when you are ever asked to present it.

If it is ever requested by a government agent or agency, the criteria for its evaluation should be fairly simple. Everyone involved in your training program should know that they have been and continue to be exposed to hazards and hazardous materials in the workplace, and what those hazards are, how to recognize them, the role Material Safety Data Sheets and labels can play in their safe handling and storage, how to read those documents, and, perhaps most important, how they are better able to protect themselves as a consequence of your training program. All this should be documented, and the records stored for a minimum of seven years.

The relevance of safety communication and hazardous material training cannot be overstated, and consequently should not be ignored.

It doesn't matter whether the threat is physical: related to personal injury or illness; or environmental: related to air we breathe, the water we drink, or the substances we are likely to find ourselves physically exposed to. Safety in the workplace is everyone's responsibility, not just the owner's or a manager's.

It doesn't matter whether we are talking about the physical danger of a fire, or about someone physically, sexually, or verbally abusive. These are real threats with real consequences, and no one is immune or exempted!

The role that training can play is critical. When executed properly, your training program can increase awareness and change behavior in a powerful and positive way by effectively and immediately moving information from those who have it, to those who need it—from where it is, to where it needs to be.

What kinds of training will you institute and how will you deliver them? Write your answers in the chart on the following page.

Thoughts

Actions

Results

CHAPTER 4

Hazardous Communication Guidelines

INTRODUCTION

This chapter will help you to become familiar with the rules applicable to hazardous communications.

In addition, it will introduce you to the established guidelines that should be followed while building a hazardous communications program, such as:

- Identifying responsible staff member(s)

It will also explain what you must both know and do to be properly prepared, with a focus on:

- Labels
- Signage
- Material Safety Data Sheets
- Employee information and training

GUIDELINES

Realistically, we've been discussing these guidelines throughout the book. Now we're going to concentrate on just exactly what these government regulations are all about and the extent of our responsibilities with regard to them.

The Hazardous Communication Standard is based upon a relatively simple principle, and that is that all workers have a right to know when and how they may have been placed at risk by the nature of the work they are doing, or by the chemicals and compounds with which they are doing it, or by the environment in which it is being done; and also to know what protective gear is available to help reduce, minimize, or eliminate their risk.

The basic premise is that the more informed and educated workers are, the more they will be able to initiate workplace polices and procedures designed to reduce exposure, establish more responsible work practices, and ultimately create a safer workplace with fewer illnesses or injuries related to chemical exposure.

There are two critical responsibilities that reside here. The first deals with the evaluation of a chemical with regard to its nature and the extent to which it is or may be hazardous, and the second is communicating that information to your employees.

Since we do not manufacture the chemicals we use, evaluation is not really something we have to worry about. The manufacturers and importers are responsible for that part of the program. However, our role is no less significant. Without trying to minimize that role, because it can often seem overwhelming, we are responsible for identifying whether or not the chemicals we are using *are* hazardous and may pose a potential danger, what chemicals they are, where to store them, how to handle them, how to dispose of them, how to document all of that, and then how to communicate it to our employees in an effective, efficient, and useful way.

FAMILIARIZATION WITH THE RULE

The federal Hazardous Communication Standard (HCS) is long, complicated, and in some places very technical in its explanations. In many cases, state standards or requirements can be even more comprehensive and demanding. They will generally, however, concentrate on two different elements: the physical nature of a hazard, such as flammability and injury from burns; and health hazards, such as irritation to the eyes or skin, damage to the lungs, or cancer.

The use of, or exposure to, virtually all chemicals carries with it at least some risk. Even the safest chemicals and compounds can become lethal if they are used irresponsibly or combined with other chemicals incorrectly. Consequently, a working knowledge of safe practices and responsible procedures is critical. Because the HCS is performance-oriented, and in order to facilitate the best possible results, the government has afforded business owners a certain latitude designed to help you customize your program for maximum results.

On the outside, it looks simple: the manufacturers, importers, and distributors are responsible for evaluating all chemicals and providing the appropriate labels and accompanying material safety data sheets. Every container must be labeled and every first shipment of a chemical must have an MSDS shipped with it. In addition, a new and updated material safety data sheet must be sent any time the chemical is modified.

If you use hazardous materials (and "use" means package, handle, or transfer—which involves virtually everyone in the automotive service industry), you must be able to document that you have a program in place that provides the appropriate information to any employees likely to be exposed.

In all questions regarding standards or regulations, always try to remember that their interpretation by government representatives will almost always fit this pattern: broad when it comes to coverage of the workers or the workplace, and very narrow and specific when it comes to your implementation.

RESPONSIBLE STAFF

The government's intention was for this legislation to result in a *process*, and not just another *program*. There is a difference, and the difference is relatively simple. A program has a beginning, a middle, and an end—a start date and a finish date.

> **pro·gram** (pro-gram) *noun* **1. a.** A listing of the order of events and other pertinent information for a public presentation. b. The presentation itself: a program of piano pieces. **2.** An ordered list of events to take place or procedures to be followed.

A process is ongoing: it, too, has a beginning ... but it isn't ever supposed to have an end!

> **proc·ess** (pros-es) *noun* Abbr. proc. **1.** A series of actions, changes, or functions bringing about a result.

In order for this process to be successful, there must be someone in charge, someone responsible for getting things moving in the beginning, someone responsible for keeping them going in order for this process to become a meaningful part of your everyday shop policies and procedures.

Accomplishing this will not necessarily be easy and is likely to require a champion, or at least a cheerleader. Your job as the owner or manager of an automotive service business will be to choose the right person for the job—someone who believes passionately that it is not only the right thing to do, but in the best interest of everyone involved. When this works, behaviors change and conditions improve: your employees will understand and support the program, they will commit themselves to its success, and everyone involved will be guided and motivated to ensure that the process is successful.

Shop Chemicals

The Hazardous Communication Standard requires a written HCS Program. It also requires as a part of that program that *all* hazardous chemicals in the workplace be listed. The list serves two purposes. First, it identifies all the chemicals that present a hazard, while establishing the scope of the potential risk to which your employees may be exposed. And second, the list itself serves to ensure that there is an MSDS on-site and available for every one of those potentially hazardous chemicals.

The best way to execute this responsibility is by actually taking an inventory of all the chemicals in the workplace. It would be wise to remember that *all* means *all*, especially when the government is involved, and that it is especially important to include every chemical regardless of its physical form, including liquids, gases, vapors, powders, fumes, and mists, whether they are contained or not. It would be equally wise to spend a few minutes thinking about the chemicals you are likely to ignore or forget about entirely, like the chemicals you clean the office or rest rooms with.

Once you have such a list, you must go through it to determine if the chemicals listed are covered by the HCS, since not all chemicals are covered.

If the chemical is not hazardous, it is not covered and should not be a part of your program.

While you are in the process of creating your list, don't forget the chemicals that might be a byproduct of any of the mechanical procedures likely to be performed in the shop, such as exhaust emissions, welding fumes, dust or debris generated while cleaning parts or components, chemical cleaning waste, contaminated liquids, etc. Identify where they are as well as what they are. In fact, a detailed map including the kinds of hazards in each location is a good way to ensure that firefighters and other rescue or law enforcement professionals know what they will be facing when they respond to an emergency.

Once you have a list of all the chemical hazards in your facility, the next step is to ensure that you have a *current* MSDS for each chemical. If you don't, call your supplier and ask for one. And document the date, time, and content of that conversation for your records. It will demonstrate good faith—that you really did, in fact, try. If your supplier is not forthcoming with the MSDS you requested verbally, request it in writing. Send it certified or registered mail, return receipt requested, to document what you did and when you did it, and keep a copy and the receipt for your records.

If you have an MSDS for a chemical that you don't have on your list or that you can't locate, you need to know why. While it may be true that you no longer have a use for that chemical, it could also be true that it might be lurking somewhere in the shop without you or any of your employees knowing where it is or even what it is. And the fact that such a chemical could be on the premises without your knowledge goes to the heart of the reasoning behind this legislation.

It is critical for you to understand the liability, both personal and professional, involved in allowing an employee to work with a chemical for which there is no safety data! It makes an emergency response to exposure to that chemical difficult, if not impossible.

PREPARATION

> *"All workplaces where employees are exposed to hazardous chemicals must have a written plan which describes how the plan will be implemented in that facility."*
>
> —*Hazard Communication Guidelines for Employer Compliance—from Appendix E (Advisory) http://www.osha.gov/SLTC/smallbusiness/sec16.html*

Just having a plan isn't likely to be good enough if a state or federal OSHA representative comes "a'calling"! He or she will want to see that your plan is actually implemented: in other words, that it is real and exists somewhere other than on the top of your desk or on a shelf in the back somewhere! This plan does not have to be complicated, but it does have to be relevant in the sense that it serves to show just how you met all the requirements involved in the program. Most importantly, your plan must reflect what is actually being done in the workplace to ensure that everyone is aware and involved when it comes to creating a safe and hazard-free environment.

There are a number of generic, one-size-fits-all plans available through our industry's trade associations and through a number of the larger program distribution groups. These are helpful at the very least, but don't think an OSHA representative will be satisfied with a plan that has no visible proof of implementation. *Your* plan must have a list of the chemicals presently in *your* facility along with the name of the person responsible for program implementation and execution. It should also include the locations of those chemicals. Everyone in the place should know that you have a program, what the purpose of the program is, how it relates to them, and where to find it, including where the written copies are and who is in charge.

You do *not* want to have an OSHA inspector come to your facility and ask program related questions only to have your employees respond with a blank stare!

If they do come, you can count on the following:

1. They will ask to see your written plan.
2. They will want to see how the plan meets the requirements with regard to labeling and posted signage.
3. They will want to see your list of on-site hazardous materials.
4. They will want to see the accompanying Material Safety Data Sheets.
5. They will want to see your training outline and proof that the training did, in fact, occur and that your employees both know and understand the material.

Labels and Signage

All distributors and manufacturers of hazardous or potentially hazardous chemicals must label, tag, or otherwise identify their products in an appropriate and approved manner consistent with OSHA guidelines as illustrated in Figure 4-1.

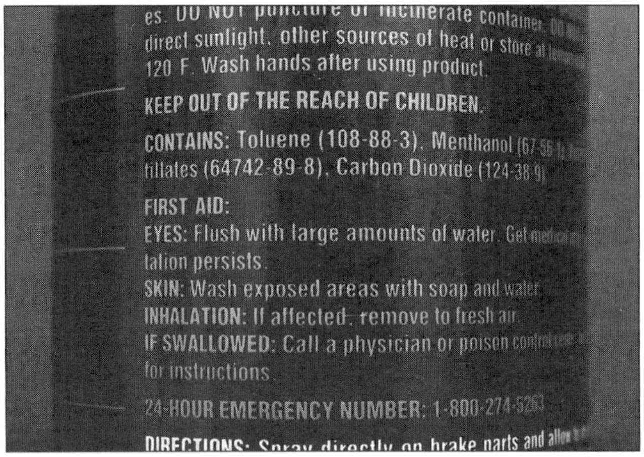

Figure 4-1 Manufacturer's hazardous material labeling.

These labels, signs, or posters should identify the material with its appropriate warnings.

> The primary information to be obtained from an OSHA-required label is an identity for the material and appropriate hazard warnings. The identity is any term which appears on the label, the MSDS, and the list of chemicals, and thus links these three sources of information. The identity used by the supplier may be a common or trade name ("Black Magic Formula"), or chemical name (1,1,1,-trichloroethane). The hazard warning is a brief statement of the hazardous effects of the chemical ("flammable," "causes lung damage"). Labels frequently contain other information, such as precautionary measures ("do not use near open flame"), but this information is provided voluntarily and is not required by the rule. Labels must be legible, and prominently displayed. There are no specific requirements for size or color, or for any specified text."

That is fairly straightforward language for a government directive, so the OSHA representative will not have any trouble asking the following questions in order to ensure that this critical part of the HCS is in place:

1. Is the person responsible for ensuring labeling of the materials brought to your shop clearly identified?
2. Is the person responsible for ensuring that anything that leaves the premises is properly labeled clearly identified?
3. Is there a description of the labeling system you have chosen?
4. Is there a description of written alternatives to labeling?
5. Is there a review process in place to update label information when necessary?

Your responsibility as an employer or as the person designated responsible for program implementation is to ensure that every chemical container entering the facility is appropriately labeled. If chemicals are moved from one container to another, the new container must also be appropriately labeled.

Realistically, it will probably remain unnecessary for you to create labels for every chemical on the grounds, since the manufacturer or distributor is already responsible for doing that. But should you have to, remember that a simple text message is adequate.

Once again, the important thing to remember here is that this is a process; and consequently, so is the responsibility. It is an on-going commitment.

Material Safety Data Sheets

Here, as with regard to labeling, it is the manufacturer's or distributor's responsibility to provide (translate as: obtain or develop) a Material Safety Data Sheet for *every* hazardous chemical they either import or produce. This must include a description of the nature of the potential hazard, its effects, its physical and chemical characteristics, and suggested protective measures.

Don't ever lose sight of the fact that the reason *they* are responsible for providing the MSDSs is because *you* are required to have them!

While there is no set format for an MSDS, there are very specific information requirements. In other words, the government does not care how the information is conveyed on an MSDS as long as all the information is there. And since you are required to have this information available for your employees, your distributors and manufacturers are required to provide this information to you. But remember, the responsibility for having a complete and comprehensive data sheet for every on-site chemical is ultimately yours, so if you receive an MSDS that is not completely filled out or if you don't receive one at all, call the person responsible and request one that will ensure your compliance.

If that doesn't work, remember that there are times when a government agency can work just as hard for you as it can against you. Call or write your local OSHA Area Office and ask for their help, and I'll bet you might be surprised at how fast an MSDS arrives at your door!

As we have already seen, there is a lot of information contained on a typical Material Safety Data Sheet. But we are primarily concerned only with the information appropriate to our immediate needs. This information is often focused on the hazardous effects of the chemicals we come in contact with in the workplace and the protective measures necessary to ensure that the recommended exposure limits are not exceeded.

As we have already seen, your employees have the right to know if or when they have been exposed to a chemical hazard or potential health risk, and the nature of that risk. The MSDSs are a response to that right, but they will not fulfill the program's requirements if no one knows what they are or where they are in the building! Consequently, it is your responsibility to see that your employees know where to find them, how to access them, and what the information provided on them is all about.

You get to decide the *where* and the *how*—and that can mean keeping them up front in a binder in the office, in the back near the stockroom (where most of the chemicals will probably be found), or on a hard drive and through a computer terminal in the shop. It doesn't matter. The only thing that matters is that they are available and easily accessible and that your employees know that.

Should you find yourself face-to-face with an OSHA investigator or representative, all you have to do is present your list of chemicals and their accompanying Material Safety Data Sheets, show that they can be accessed instantly, and that you have:

- Designated someone to be responsible for obtaining and maintaining the MSDSs
- Created a system for maintaining them (e.g., in notebooks or in a computer), and enabling your employees to access them

- Developed a procedure detailing what to do when the MSDS is not received at the time of the first shipment
- Described alternatives to the actual data sheets in the workplace, if applicable or acceptable

As far as the Occupational Safety and Health Administration is concerned, the most important aspect of the written program with regard to the MSDSs is responsibility—who will be responsible for obtaining and maintaining the MSDSs for each and every hazardous chemical on site.

EMPLOYEE INFORMATION AND TRAINING

The key with regard to employee information and training is to recognize that it applies to every employee who might be exposed to the risk of hazardous chemicals in the workplace, and that the information and training must be furnished prior to initial exposure and/or when the hazard changes in any way. In plain English, that means that there is no excuse for someone being exposed unknowingly. Period!

"Exposure" or "exposed" under the rule means that:

> "an employee is subjected to a hazardous chemical in the course of employment through any route of entry (inhalation, ingestion, skin contact or absorption, etc.) and includes potential (e.g., accidental or possible) exposure."

This information and training can be based either upon the individual chemicals found in the workplace (generally acceptable if there only a few chemicals to deal with), or based upon hazard categories (e.g., flammable liquids, corrosive materials, carcinogens).

This is a critical part of your overall program. The fact that there is information regarding these hazards, along with the appropriate protective measures, provided on the written labels and material safety data sheets is no guarantee that your workers will understand what they are, what they mean, or what to do. The only way that kind of knowledge is likely to occur is through effective information and training. That is the only way you will be certain that they—the people who work both for you and with you—can read and understand the information presented there, determine where both the necessary information and the description of risks can be found, how this information can best be utilized in the pursuit of a risk- and hazard-free workplace, and recognize the risks of exposure to the chemicals they might come in contact with, along with a number of ways to protect themselves.

SUMMARY

The Hazardous Communication Standard is all about the responsibility for providing workers with the education and information they need to help create a safer workplace, resulting in fewer illnesses or injuries related to chemical exposure or any other kind of hazard.

The focus is on the evaluation of a chemical with regard to its nature and the extent to which it is or may be hazardous, and communicating that information to everyone in the organization. Your responsibility as the owner or manager is to ensure that everyone in the organization recognizes the physical nature of both workplace hazards, such as flammability and injury from burns; and health hazards, such as irritation to the eyes or skin, damage to the lungs, or cancer.

There is risk involved in the use of virtually all chemicals. Even the safest chemicals and compounds can become lethal when used irresponsibly or combined incorrectly. Consequently, a working knowledge of safe practices and responsible procedures is critical. The government has afforded business owners a certain degree of latitude when it comes to customizing your program for maximum results. If the use of hazardous material is in any way a part of your everyday business operations, you must be able to document that you have a program in place that provides the required information to any employees likely to be exposed. You must also ensure that there is someone on staff designated to accept and administer these responsibilities—someone to answer the questions, someone to document and deliver the training. And finally, you must not only have a plan—you must have the documentation to support the fact that everyone in the organization knows where that plan is kept and understands what it is all about.

In addition, you must label and identify all the hazards associated with your workplace, support those labels with Material Safety Data Sheets, and reinforce them with the proper training.

Remember—when it comes to coverage of the workers or the workplace, the government's interpretation of the law will be very broad, but when it comes to the execution of your written plan, it will be very narrow and specific.

Are you aware of the chemical or safety hazards in your workplace or school environment? Do you know where the written documentation is kept? Where are the Material Safety Data Sheets stored? Who is identified as the person responsible for the program or the training in your workplace?

Take a moment to think about how these standards apply to you, and write your answers to these questions in the chart on the following page.

Thoughts

Actions

Results

CHAPTER 5

Personal Protective Equipment

INTRODUCTION

This chapter begins with an introduction to hazard assessment and the need to know what kind of personal injuries can be associated with hazards in the workplace and what can be done to reduce the risk of:

- Hazards to the eyes and face

It then goes on to explore the personal protective equipment necessary to secure a safe and accident free environment, especially with regard to:

- Ear protection
- Respiratory protection
- Torso protection
- Arm and hand protection
- Foot and leg protection

HAZARD ASSESSMENT

You and I will be spending a fair amount of time in this chapter, if for no other reason than my personal belief in its particular relevance. You see, as I mentioned in the introduction to this book, I grew up in a pre-OSHA automotive industry and the scars that lack of awareness and concern left as a testament to the importance of safety communication have made me a passionate advocate of safety in the workplace.

I have had metal particles removed from both eyes, as well as metal shards and slivers removed from other parts of my body. And as a result, I cannot utilize MRI technology in diagnosing any future health problems because of the danger of any remaining metal particles migrating through my body as a result of the intense magnetic field. I have a high frequency hearing loss resulting from a combination of thirty-plus years using air tools that weren't designed to regulate noise level and my failure to protect myself from the resulting noise pollution. I have slipped and fallen because of wet and slippery surfaces on the shop floor; bumped my head against a hoist or working under a dashboard; suffered an assortment of cuts, abrasions, and lacerations while installing or removing any one of a hundred different components and assemblies while working on a car or a truck; and just generally disregarded sound, responsible, safety practices while trying to finish a customer's vehicle on time.

I was not unnecessarily cavalier about safety in the workplace, nor was I more of a klutz than anyone else I knew working in automotive service at that time. The fact of the matter was, we just didn't know any better. Regardless, looking back, it wasn't worth it. The price was too high.

Consequently, it is very important to me that you understand exactly what your responsibilities as a business owner or manager, and/or as an employee, are. As the business owner and/or manager, it is your responsibility, your job, to determine if hazards that require the use of personal protective equipment are present, or likely to be present in your work environment. If in fact there is a possibility such hazards can, may, or actually do exist, you must select the hazards that present the most clear and present danger and have those employees engaged in activities that may put them at risk use "properly fitted personal protective equipment suitable for protection from these hazards."

It is also your responsibility to certify in writing that a workplace hazard assessment has been performed and that the appropriate steps to mitigate any workplace hazards have been taken.

Furthermore, the OSHA standard pertaining to these particular issues specifically prohibits the use of defective or damaged personal protective equipment. This basically means that if one of your employees is injured and the personal protective equipment they were wearing (if they were wearing any at all) was not what it should have been, and consequently did not provide the level of protection necessary to keep that employee out of harm's way, you won't have to look further than your mirror to find the individual most likely to be held accountable.

Training

Perhaps one of the most important training issues we can discuss is the educational process necessary to ensure that an employee is able to recognize when it might be appropriate to utilize personal protective equipment in the first place; which type of equipment would serve to best protect the employee; how the equipment should be worn; the limits of the protection that particular piece of personal protective equipment provides; its proper care and maintenance; its useful service life; and how to safely dispose of that equipment after its useful service life is over.

In cases where more than one type of personal protective equipment is available and can provide adequate protection, both the shop owner/manager and/or the technician should be able to choose the equipment they are the most comfortable using.

But it's not enough to merely present the training. An employer must certify in writing that the appropriate training has been carried out and more than that, the employer must certify in writing that the employee understood it and is able to apply what she has learned. This certification as well as the record of the actual training should include the name of each individual being trained, the date(s) the training occurred, and a description of the training material itself.

Head Protection

You probably haven't seen many hard hats used in many of the automotive service facilities you may have visited over the years, but that doesn't mean that wearing one isn't a good idea. Protecting everyone from the possibility of a serious head injury should be an important factor in any safety program. Statistics confirm that most workers who suffered serious head injuries were not wearing the appropriate protective equipment, meaning head protection such as a helmet. Perhaps more important, the majority of these workers were injured while performing their normal jobs at their regular worksites—which basically means that they were not doing anything exotic, different, or out of the ordinary.

Research conducted by the Bureau of Labor Statistics (BLS) indicated that the greatest number of accidents involving head trauma occurred where *employers* failed to require employees to wear the appropriate protective head gear. In that same survey, most of the employees surveyed indicated they were satisfied that such protective equipment was not only *practical*, it was *necessary*.

As uncomfortable as the conclusion may be, an argument can be made that since serious injury *did* occur on the job, and since it is the owner's responsibility to either provide a "safe and hazard-free environment" or make every effort to provide the equipment necessary to protect employees from the hazards that do exist, the owner failed to meet OSHA standards, and is therefore accountable.

I can honestly say that I'm not sure I know any automotive repair professionals who have not had the hood supports fail on a vehicle while they were working under the hood and had that hood come crashing down on them. I don't know very many technicians who have not at one time or another become so engrossed in what they were doing under the vehicle that they became momentarily distracted or disoriented, turned quickly without thinking or looking, and managed to bump their head against a vehicle's hood, a trunk lid, the lift itself, or some other moving or stationary object. So, whether or not head protection is made available by the shop owner, or deemed necessary by the shop manager, or even utilized by the technicians themselves if provided, it would be hard to argue that wearing such equipment is anything but prudent.

Protective gear designed to help guard against such hazards should resist puncture in addition to softening the shock of the blow. However, neither is likely to help very much if this type of personal protective equipment is not required by the employer or utilized by the person at risk of being injured.

Eye and Face Protection

The same Bureau of Labor Statistics study found that about 60 percent of workers who suffered eye injuries were not wearing eye or face protection at the time of their injury. The standard answer when asked why the appropriate protective gear was not being used at the time of the accident was that the use of such equipment was not required or even customary for that particular workplace or profession, or normal for the kind of work being done. As someone who has personally experienced the inconvenience, pain and discomfort associated with not wearing eye or face protection, and the subsequent injuries that followed, this is an issue I can address from the perspective of personal experience. And because of those self-inflicted injuries, and the trauma of having metal particles removed from both eyes at different times in my career because I didn't wear the kind of protective equipment illustrated in Figure 5-1, it is also something I am quite passionate about.

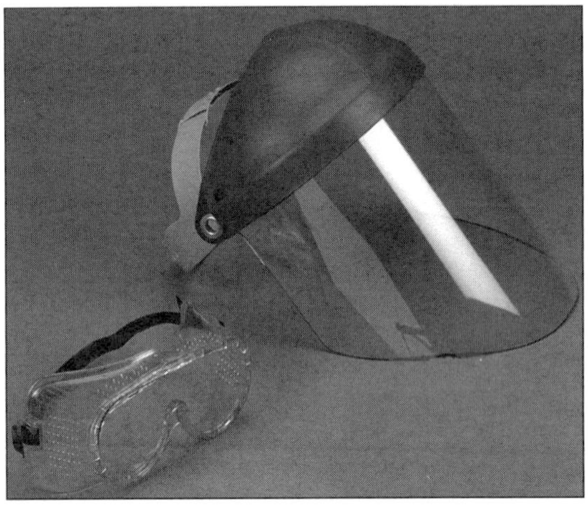

Figure 5-1 Eye and face protection.

However, my feelings are irrelevant, despite the passion I bring to this particular area of discussion, because the use of such protective gear is the law.

> "Suitable protection must be provided where the possibility of eye or face injuries from flying particles, molten metal, liquid chemicals, acids or caustic liquids, chemical gases or vapors, potentially injurious light radiation or a combination of any of these is likely or even possible."

Any protective equipment must be designed with the specific hazards being guarded against in mind—otherwise that equipment will prove useless. That means it should fit properly and remain comfortable while it is being worn or used under the conditions for which it was designed, or the people who need that protection the most are likely to wear it least. The equipment should be easily accessible, durable, clean, capable of being disinfected if necessary, well maintained, and in good working order.

Certain kinds of work, such as gas or electric welding, fabrication or cutting that utilizes an oxygen and acetylene cutting torch, or any other work that exposes the worker to what OSHA refers to as "injurious light radiation," should involve the use of eye protection designed to filter out dangerous radiation with a shade number appropriate for the work being performed. The following tables in Figures 5-2a and 5-2b list the appropriate shade numbers for various work operations.

Filter Lens Protection Against Radiant Energy

Operation	Electrode Size (1/32 inch diameter standard)	Arc Current (Amps)	Minimum* Protective Shade
Shielded metal arc welding	<3/32 3/32-5/32 r/32-8/32 >8/32	<60 60-160 160-250 250-500	7 8 10 11
Gas metal arc welding an flux cored arc welding		<60 60-160 160-250 250-500	7 10 10 10
Gas Tungsten arc welding		<50 50-150 150-500	8 8 10
Air carbon arc cutting	(Light) (Heavy)	<500 500-1000	10 10
Plasma arc welding		<20 20-100 100-400 400-800	6 8 10 11
Plasma arc cutting	(light)** (medium)** (heavy)**	<300 300-400 400-800	8 9 10
Torch brazing Torch soldering Carbon arc welding		-- -- --	3 2 14

* *As a rule of thumb, start with a shade that is too dark to see the weld zone (the darkest lens carries a value of 10). Then go to a lighter shade which gives sufficient view of the weld zone without going below the minimum. In oxyfuel gas welding or cutting where the torch produces a high yellow light, it is desirable to use a filter lens that absorbs the yellow or sodium line in the visible light of the (spectrum) operation.*

** *These values apply where the actual arc is clearly seen. Experience has shown that lighter filters may be used when the arc is hidden by the workpiece.*

Fig 5-2a Filter lens protection.

Operation	Plate Thickness: Inches	Plate Thickness: Millimeters	Minimum Protective Shade
Gas welding:			
Light	<1/8	<3.2	4
Medium	1/8 to ½	3.2 to 12.5	5
Heavy	>1/2	>12.7	6
Oxygen Cutting:			
Light	<1	<2.5	3
Medium	1 to 62	5 to 150	4
Heavy	>62	>150	5

Fig 5-2b Plate thickness.

Eye Wash

The official policy of OSHA with regard to eye protection, in virtually all cases, is that the employee's eyes should be protected from injury by the use of proper personal protective equipment. This is especially true when it comes to the handling of potentially dangerous or hazardous chemicals. Since the use of appropriate eye protection is the primary defense against such injuries, the use of a liquid eye-wash or face-wash is a secondary or fallback method of reducing the consequences of injury caused by chemical exposures that involve the eyes. Nevertheless, to be effective, an eye-wash or face-wash station, or sterile liquid eye-wash materials placed in a first aid cabinet or station(s), should be easily accessible throughout the shop and capable of being self-administered. First-aid instructions should be posted close to potential danger spots, since any delay in immediate aid or an early mistake in dealing with an injury can result in lasting damage. The ability to wash the face and the entire area around the eyes, where additional chemical is often present, with a constant flow of safe liquid can provide additional benefit, at least until the employee is able to seek care from an appropriate source.

Guidelines

OSHA guidelines for emergency eye-wash and shower equipment generally fall into two categories. The first deals with general regulations and the second with more specific regulations. The general regulation for emergency eye-wash and shower equipment states:

> "Where the eyes or body of any person may be exposed to injurious corrosive materials, suitable facilities for quick drenching or flushing of the eyes and body shall be provided within the work area for immediate emergency use." [29 CFR 1910.15]

This applies to all facilities that require the installation of emergency eye wash and shower equipment as a form of first aid, and generally includes automotive service facilities like yours and mine. The second type of regulation is specific to other industries, including those utilizing open surface tank storage and handling of anhydrous ammonia; powered industrial lift trucks; pulp, paper, and paperboard manufacturing; telecommunications; and others. Like many government regulations, they are quite specific about where and when emergency eye wash and shower equipment must be made available, but aren't at all specific when addressing minimum operating, installation, or set-up requirements. The American National Standards Institute

(ANSI) developed ANSI standard Z358.11990 to help select and install emergency equipment that meets or exceeds the OSHA requirements.

The following specifications split the two types of primary eye wash units and/or shower equipment into two basic categories. The first category includes plumbed eye wash units, which are described as those permanently connected to a source of potable water—in other words, your internal plumbing, as illustrated in Figure 5-3.

Figure 5-3 Eye wash station.

And the second category, described as gravity feed eye wash devices, are self-contained and include their own flushing fluid, which must be replaced or refilled after use.

To give you an idea of just how specific these specifications can be, and just how dangerous *not knowing what you don't know* can be, I have included the following as an example. It relates to the ANSI standards mentioned above for both plumbed and gravity fed eye wash devices.

Here are the specifications for plumbed devices suggested by ANSI, complete with the legal disclaimer that accompanies them:

1. Heads must be positioned 33 to 45 inches from the floor and 6 inches from the wall or nearest obstruction. They must have the capacity to deliver 0.4 gallons of fluid per minute for 15 minutes.
2. Valves should activate in 1 second or less and stay open, leaving the hands free.
3. The unit should be installed within 100 feet or 10 seconds away from the hazard. (Consult a medical professional to determine the appropriate distance for harsh acids and caustics. Higher hazards require closer distance.)
4. Maintenance and training is required for eye wash equipment. Plumbed units shall be activated weekly to verify proper operation. Gravity feed units shall be maintained according to the manufacturer's instructions. All employees who might be exposed to a chemical splash shall be trained in the use of the equipment.

Eye/Face Wash and Drench Hoses

An eye/face wash device is used to irrigate and flush both the face and the eyes. Specifications for these units are similar to the eye wash. Heads should be positioned 33 to 45 inches from the floor and 6 inches from the wall or nearest obstruction. Large heads to cover both eyes and face or regular size eye wash heads plus a face spraying ring should be provided. The device should deliver 3 gallons of fluid per minute for 15 minutes. Valves, installation, maintenance, and training should be the same as for eye wash units as outlined above. A drench hose is a flexible hose connected to a water supply and is used to irrigate and flush eyes, face, and body areas.

Specifications:

1. The head must supply 3 gallons of water per minute.
2. The valve must activate in 1 second or less.
3. The device must be assembled according to the manufacturer's instructions and located in a well-lit area with a sign identifying it.
4. Maintenance and training should be the same as for eye wash units.

It should be noted here that ANSI suggests that while held drench hoses can support shower and eye wash units, they should not be considered a replacement for them.

A supplementary eye wash unit can support plumbed units, gravity feed units, or both, by delivering immediate flushing fluid. Personal units located near work stations can provide immediate flushing if needed. This equipment doesn't have to meet the requirements of plumbed or gravity feed units and can't be substituted for them, but personal units can support plumbed or gravity feed eye wash units. Important: All primary emergency eye/face wash units must have a minimum 15-minute uninterrupted supply of clean water.

Information in this update is intended for general information purposes only. This is not a substitute for review of the applicable government regulations and standards, and should not be construed as legal advice or opinion.

—April 1996, Todd Hamilton, NAEDA Safety Consultant/Federated Insurance, from NAEDA Equipment Dealer, *www.omeda.org/fastfacts/5210.htm*

Selection

Selecting the appropriate type of eye, face, or face-and-eye protection should be based upon the specific nature of the hazard and the ability of the protective gear to mitigate or protect against the hazard or danger for which it was designed. Where there is a choice of different kinds of protective gear, and the degree of protection provided is equal or close to equal, the degree of comfort may be used as the deciding factor. The Bureau of Labor Statistics survey that we have been referencing indicated that few workers ever complained about poor vision or discomfort with regard to the personal eye protection equipment they were wearing.

I should point out here that the use of corrective lenses or prescription glasses not specifically designed to meet or exceed OSHA requirements is not a substitute for the appropriate face or eye protection; and that those required by OSHA standards to wear such protective gear must wear face shields, goggles, or spectacles:

- Spectacles with protective lenses and side shields providing optical correction
- Goggles worn over corrective spectacles without disturbing the adjustment of the spectacles
- Goggles that incorporate corrective lenses mounted behind the protective lenses

The bottom line here is to find gear that you and/or your employees will be comfortable using and then set the example by using it yourself!

Inspection and Maintenance

No one should have to tell you that keeping the lenses of any kind of eye protection equipment clean is essential. It is no less important than taking care of any tool or piece of equipment essential to the performance of your responsibilities as a professional in your trade. It is just as critical as making sure your chisels are sharp with ends that aren't mushroomed, or that your air tools are well lubricated. Aside from that, dirty lenses can impair your vision and impaired vision can result in accidents. Continuous vision through cloudy or dirty lenses can cause eye strain, and eye strain can often result in fatigue, discomfort, and/or headaches—all of which can serve as an excuse for not wearing the eye protection.

While cleaning and inspecting this kind of protective gear every day is recommended, it may not be realistic. It is more important to ensure that this kind of protective gear is inspected on a *regular* basis—however frequent you decide "regular" is—and that damaged, pitted, or scratched lenses or face coverings, and/or worn or damaged frames or shields are replaced when they are no longer safe or useful.

The same kind of common sense care should be given to the eye protection you use in the shop as that used when handling prescription lenses. This means that frames that no longer fit properly or headbands that no longer hold the eye protector in the proper position should be replaced with protective gear that fits. And personal protective gear in general use—equipment used by anyone and everyone in the shop—should be disinfected both before and after being used, both for safety and for hygiene.

There are several ways you can disinfect eye-protective equipment. The most effective method is to thoroughly clean all parts with soap and warm water, carefully rinse away all traces of soap, and then immerse all parts in a solution of germicidal deodorant fungicide for ten minutes. After ten minutes, you can remove all the parts from the solution and then suspend them in a clean place for air drying at room temperature or with heated air. It is recommended that you do not rinse these parts after removing them from the fungicidal solution, because this will remove the germicidal residue which retains its effectiveness after drying.

Afterward, the dry parts or items should be placed in a clean, dust-proof container, such as a box, bag, or plastic envelope, to protect them until they are used again.

EAR PROTECTION

Have you ever measured just how much background or surface noise you are exposed to? I have a simple experiment for you. All of us are creatures of habit. We generally do the same things in the same order every time we do them, if for no other reason than that these rituals have become familiar to us. The next time you are finishing up and getting ready to go home—before you close anything or shut anything off—I would like you to do the following for me: take a moment at the end of each step and just listen. Stop and listen when you close the shop doors before you close the compressor for the night. Stop and listen after you drain the compressor and shut it down. Stop and listen to the level of noise before you close the lights in the shop, particularly if they are fluorescent fixtures, and listen to the noise—or the absence of noise—left when the ballasts for those fluorescent fixtures are finally quiet. And then do the same at each step along the way as you get ready to walk out the door.

This is not some foolish, frivolous request that has no relation to reality. It is in fact an experiment I execute on a regular basis myself, just to ensure that I am not exposed to any more noise than is absolutely normal or necessary. I think you will be amazed at just how much noise you are exposed to—noise that you may not think about or even be aware of, for that matter.

Background noise is distracting, and it can be dangerous. It can cause physical and psychological stress without your ever being aware of your exposure to it. Exposure to high noise levels can cause hearing impairment or even, in extreme cases, hearing loss. This is permanent, irreversible damage for which there is no medical remedy. Consequently, the only way to avoid this kind of hazard is either to control the exposure or to control the noise.

Because this is one of the hazards specifically addressed by OSHA, it cannot and should not be ignored. Where specifically designed ear protection is required, it is your responsibility as the owner or manager to see that it is used. It may be more important, however, to demonstrate to your employees just how important this kind of protection can be, as well as to explain the life-long consequences of not taking such a threat seriously. There is really no excuse.

RESPIRATORY PROTECTION

If you walk into any automotive service facility in the country and spend any amount of time in the shop itself, you won't need OSHA to alert you to the fact that there are airborne chemicals that could be considered hazardous to your personal health or safety. Detergents, cleaning agents, fuel, oil, anti-freeze, hydraulic fluids of all kinds, friction material residue from either brakes or clutches, and vehicle tailpipe emissions are just a few of the toxic materials those of us engaged in automotive service are exposed to every day.

In too many cases, these vapors and airborne particulates are known carcinogens—chemicals and compounds known to cause cancer. Consequently, their responsible use, handling, storage, and disposal are critical, just as using, handling, storing, or disposing of them irresponsibly is dangerous.

We talked about exposure limits in Chapter 1 when we were discussing the information contained on a Material Safety Data Sheet with regard to Occupational Exposure Limits (OEL) and Maximum Exposure Limits (MEL). OSHA has established standards to protect you and those around you from exposure to such hazards in terms of Permissible Exposure Limits, or PELs. OSHA sets these regulatory limits to protect workers against the health effects of hazardous substances in the air or on the surface of your body. These limits are based upon an eight-hour time-weighted average (TWA) exposure to certain chemicals and compounds.

Despite the nature of the chemicals and compounds we are exposed to daily, in most cases the concentration limits are not sufficient to require the use of respirators by statute. Consequently, while respirators are commonly used in the collision industry, as a rule they are not as common in the service industry. This may have more than a little to do with the physical space and the open nature of most repair shops, as opposed to the controlled, almost closed environment found in most paint and body shops.

Nevertheless, respirators are appropriate where exposure levels exceed the permissible exposure limit (PEL). To determine if or when you are affected, you must first inventory the potentially dangerous or hazardous chemicals or compounds in use in your facility and then determine exactly what the appropriate PELs are.

However, the fact that there may not be an appropriate standard for some of the chemicals you are exposed to is no reason *not* to use personal protective gear designed to filter the air you are breathing from potentially dangerous substances! Common sense, instinct, and intuition should tell you that if it's hard to breathe in without coughing or choking, the kind of protective equipment illustrated in Figure 5-4 is appropriate!

Figure 5-4 Breathing masks.

TORSO PROTECTION

There are countless hazards within the automotive service environment that can threaten the body. There is a constant danger of fire, resulting from the combustion of flammable materials, and of the heat associated with that fire. There is always the possibility of contact with superheated liquids that can result from a ruptured cooling system or transmission line, as well as the normal hazards associated with working on any machine that is itself operating at high speed and high temperature while you are working on it.

There is appropriate protective equipment designed to offer at least some degree of protection from any and all of these dangers and/or hazards, and as a result of the range and availability of these solutions or potential solutions, there are choices. As an example, wool and specially treated cotton are two natural fibers that are fire-resistant and comfortable, since they adapt well to changing workplace temperatures—while other fabrics may present as much of a hazard as the fire itself. There are closely woven cotton fabrics that are actually very effective at protecting the body from lacerations or abrasions; heat-resistant materials, such as leather, that are often used in protective clothing to guard against dry heat and flame; and rubber or rubberized fabrics, neoprene, and plastics that can offer reasonably good protection against a number of acids and other dangerous chemicals.

Just remember that it is your responsibility as an owner or manager to provide your employees protection from injuries associated with the performance of their job, especially with regard to:

- Intense heat
- Burns resulting from contact with hot or super-heated liquids
- Cuts, lacerations, or abrasions
- Contact with hazardous chemicals
- Contact with potentially infectious materials
- Radiation
- Impact from tools, machinery, and materials

ARM AND HAND PROTECTION

Virtually the same hazards that would apply to protection of the body also apply to the protection of the arms and the hands. And realistically, these extremities are generally at risk of serious injury far more often in our environment. Many of these risks or hazards are associated with the use of machines, tools, and/or various other pieces of equipment; and in a number of cases, the point-of-operation machine guards that are an integral part of the equipment, designed to protect the operator from unnecessary injury, are sufficient. However, in order for these point-of-operation guards to work, they must be installed, in good condition, and most of all, they must be utilized.

I've been in too many shops where these point-of-operations safeguards have been disabled, subverted, or removed entirely. How many shops have you been in where the eye shield or protective guard was removed from a bench grinder? Too many, I would guess.

Other examples of injuries to arms and hands can result from burns, cuts, electrical shock, amputation, and/or the absorption of chemicals or dangerous compounds. There is a wide assortment of protective equipment appropriate for each potential hazard, and there are gloves, hand pads, sleeves, and gauntlets designed to protect the worker from them.

It is your responsibility as an employer or a manager to determine what kind of protection is appropriate for the kind of work being performed. Choices will depend upon the degree of dexterity required; the duration, frequency, and degree of exposure to hazards; and the physical stresses the employee will be exposed to.

In order to meet this responsibility effectively, it is critical to understand the performance characteristics of the protective gloves you choose relative to the specific hazard anticipated—e.g., exposure to chemicals, heat, or flames. Before purchasing any gloves to protect an employee from a specific hazard or hazards, you should insist that the gloves meet the appropriate test standard(s) for the hazard(s) anticipated, and that the manufacturer or supplier document their performance. For example, for protection against chemical hazards, the nature of the chemical or chemicals being handled must be determined—particularly, their ability to penetrate the various kinds of materials most protective gloves are made from. There are many chemicals that you will come in contact with in the automotive service environment that will pass directly through a host of different glove materials to contact the skin, with the clear and present danger of the adverse systemic effects that may accompany absorption.

The protective gear or equipment should always be selected with a critical consideration of the job that will need to be performed.

FOOT AND LEG PROTECTION

As with most of the other accidents cited in the BLS survey, most of the workers who suffered injuries to their feet or their legs were not wearing the appropriate personal protective equipment—equipment that could have diminished the severity of the injury and saved the injured worker pain, suffering, and economic hardship. Most of these injuries occurred while workers were performing job activities normal to their profession, and in almost all cases the workers were not required to wear safety boots or shoes by their employers.

In order to identify whether or not foot or leg protection is appropriate—in other words, whether or not such a hazard exists in *your* workplace—you are almost compelled to perform a *safety audit*: something we discussed in greater detail in Chapter 2. However, for now, I would simply like to point out that if there is a *possibility* that one of your employees could be injured from falling or rolling objects; punctures, cuts, or lacerations from sharp objects; contact with molten metal, hot surfaces, or wet slippery surfaces; then those workers should be issued adequate protection in the way of foot guards, safety shoes, and/or boots and leggings (leggings protect the lower leg and feet from molten metal or welding sparks).

Personal protective equipment is available as aluminum alloy, fiberglass, or galvanized steel foot guards which fit over conventional work shoes or boots; and/or safety shoes or boots that have a built-in impact-resistant toe.

SUMMARY

As someone who has too often recognized the potential benefit of personal protective equipment *after* a visit to the local emergency ward for failing to use it—in other words, too late!—you can take it from me when I tell you that its importance cannot be overstated!

It doesn't matter what kind of personal protective equipment we are talking about. The use of ear protection, gloves, gauntlets, leggings, a hardhat, or an apron can save you the pain and discomfort of a chronic condition or a long-term recovery.

Think about the different kinds of personal protective equipment there are, the benefits associated with using PPE and the consequences of not using it, what you have and what you might need, and then list all these in the chart on the following page.

Thoughts

Actions

Results

CHAPTER 6

Material Handling and Storage

INTRODUCTION

This chapter will introduce you to the hazards of handling and storing hazardous materials:

- The how and the why of it
- Policies and procedures
- Controlling inventories
- Containing spills
- Emergency planning

THE HOW AND WHY

Without belaboring the point, hazardous wastes and materials are, by definition, hazardous. That means they constitute a threat to those individuals who might be exposed to them, especially if they are unaware of the risks involved in the handling or use of such products or materials.

Remember, we are talking about substances that are ignitable, corrosive, reactive, or toxic.

ig·nite (ig-nīt) *verb* ig·nit·ed, ig·nit·ing, ig·nites verb, transitive **1. a.** To cause to burn. **b.** To set fire to. **2.** To subject to great heat, especially to make luminous by heat.

cor·ro·sive (kə-rō-siv) *adjective* 1. Having the capability or tendency to cause corrosion: a corrosive acid. 2. Gradually destructive; steadily harmful.

re·ac·tive (rē-ak-tiv) *adjective* **1.** Tending to be responsive or to react to a stimulus. **2.** Characterized by reaction. 3. Tending to participate readily in chemical or physical reactions.

tox·ic (tok-sik) *adjective* 1. Of, relating to, or caused by a toxin or other poison. 2. Capable of causing injury or death, especially by chemical means; poisonous.

And we shouldn't lose sight of exactly what those four terms mean to us.

Recognizing that there is a right way and a wrong way, an appropriate way and ways that are less appropriate, we must identify the *best* way to distinguish one type of hazard from another (through the use of icons as illustrated in Figure 6-1), along with the *best* ways to handle and store such wastes and materials—the best way for us, especially with regard to the safety of everyone involved, as well as the best way for the environment.

Figure 6-1 The use of icons for quick identification.

POLICIES AND PROCEDURES

Throughout this text we've been talking about a number of different issues that really must become an integral part of your company's culture. The only way I know to make that happen effectively and efficiently is to build the desired behaviors into your everyday policies and procedures, and then *live* those policies and procedures until they become habit—the natural behavior of everyone in the company.

This would be important just by virtue of the subject matter itself, but it becomes especially critical because it is mandated by state and federal law. Consequently, it is your responsibility to see not only that everyone has the information they need to remain safe in a risk-free or at least a risk-reduced environment, but also that they understand the information they have. That means taking everything we've been talking about throughout this book and integrating it into your shop's policies and everyday procedures. It means continual reinforcement though continual review. It means constant training and education. And it means continually looking for better ways to minimize the inherent risks involved in doing what we do.

Examples

Your policies and procedures should follow local and federal requirements, so be sure to determine exactly what those requirements are. However, by way of example, let's take a quick look at some of the concerns and considerations necessary to build an appropriate set of policies and procedures for something as basic as handling a liquid waste like used coolant or anti-freeze—something that might be appropriate for all of us. We'll use engine coolant as an example because it is something almost universal to the industry, but we could just as easily have chosen cleaning solvents, refrigerant, waste oil, or a half dozen other potentially dangerous chemicals.

Engine coolant (ethylene glycol-based coolant especially) is considered to be moderately toxic for a number of reasons, not the least of which is the presence of a corrosion inhibiting chemical known as triethanolamine, known to cause cancer in laboratory rats. Coolant removed from the radiator of a passenger car or truck after the vehicle has been in service for any length of time is considered a hazardous waste across most of North America because of the high concentrations of lead (in lead-soldered radiators) and other contaminants which can cause serious health side-effects if present in sufficient quantities. Consequently, it must be handled as such. That generally means it must be stored on a non-porous surface which will not allow any spilled waste to enter the ground water system, as illustrated in Figure 6-2, and it should be clearly identified as a hazardous waste by the use of both signs and labels.

Figure 6-2 Waste anti-freeze storage.

If stored outside the building, most states will demand that it be stored within a secondary containment device designed to hold 110 percent of the quantity of the primary containment vessel. In other words, secondary containment for a fifty-five gallon drum would be just over sixty gallons. Regardless of where it is stored, nothing but the labeled material identified as being inside that container should be placed in the container.

Waste oil, oil removed from the crankcase of a passenger car or truck, should be handled in much the same way. I'd like to share with you just one section of the MassBay Community College, Science and Advanced Technology Institute, Automotive Technicians Environmental Action Movement (ATEAM) curriculum sent to me as research material by Chuck Pearson, Special Staff Assistant, Technology Center. If you take a good look at this material, you can see just how easily it could be integrated into a shop policies and procedures manual. Let me paraphrase the "Best Management Practices" as follows:

1. Drip pans or other portable containment devices should be used to avoid having to use absorbents to clean up any oil spilled during oil changes.
2. The goal should be an oil and filter service completed without spilling any oil.
3. The appropriate waste storage area should be designated and identified and waste oil should be segregated from other liquid wastes.
4. Waste oil should be labeled as both a "Hazardous Waste" and "Waste Oil".
5. Oily debris resulting from cleanup should be kept to a minimum.
6. No mixing of other chemicals with waste oil should be allowed. However, most other lubricants such as transmission oil and gear lube are generally defined as "oil" and can generally be added to the used oil container. Nevertheless, it is your responsibility to check with local government agencies in order to be sure of the regulatory requirements in your area.

With regard to disposal options:

1. Waste oil should be transported by a licensed (and bonded) hazardous waste carrier to an approved oil recycling facility, and a hazardous waste manifest should be used to record the shipment.
2. Waste oil could be recovered by burning it as heating oil in an approved waste oil-fired heater.

Now let's take a more specific look at what the curriculum at MassBay has to say about what we need to know and what we should be doing about the used oil filters we remove from the vehicles we service every day. Pay close attention. There are a number of policies and procedures that should jump right out of the text at you.

> Health and Safety Issues:
>
> Oil filters are designed to remove particulate matter from the oil. Both lead and benzene, known toxics, have been found in spent filter paper. To avoid skin contact, workers should wear impervious clothing and gloves during handling. Spent oil filters also contain significant amounts of free-flowing oil, which is classified as a hazardous waste in Massachusetts.

The first policy and procedure might be the use of "impervious clothing and gloves during handling." The second is set up by identifying the oil trapped within the filter as a hazardous waste (in Massachusetts).

> Regulatory Issues:
>
> Waste oil is a regulated hazardous waste in Massachusetts. Therefore, DEP encourages the recycling, or processing, of used oil filters and establishing a management approach. To meet the basic federal and state requirement, technicians who change oil must puncture and drain the removed filter for a minimum of 12 hours.
>
> The filter can then be managed as a solid waste, rather than a hazardous waste, but all free-flowing oil must be managed as a hazardous waste.

The second and/or third policy and procedure elements suggest that while the filter itself can be considered a solid waste and can be treated as such, the oil trapped within the filter is a hazardous waste and must be drained for at least twelve hours after the shell of the filter has been punctured.

Below, there are three additional policy and procedure elements for your consideration.

> Best Management Practices:
>
> The engine should be at operating temperature before performing an oil and filter change service. The used oil filter contains 12 ounces or more of used oil. To remove as much of this as possible, the technician must select one of the following practices:
>
> 1. After removal, puncture the filter anti-drain back valve or the filter dome end and "hot drain" for a minimum of 12 hours into a used oil collection drum. At the end of this period the filter may be discarded in the solid waste trash.
> 2. After removal, "hot drain" and crush the filter, allowing the free-flowing oil to drain into a used oil collection drum. The filter may then be discarded in the solid waste trash.
> 3. After removal, "hot drain" and dismantle the filter(typically the drained filters will be shipped in a 55 gallon drum to an off-site processor who will mechanically separate the components into scrap metal, filter element, and recovered oil).

Option number 3 is considered to be the most environmentally responsible method. The company that is removing the used waste oil will "backcharge" the generator if the used waste oil is found to contain significant amounts of other chemicals.

INVENTORY CONTROL

Monitoring and managing all aspects of a hazardous communication program will often impact everyone in the workplace in a positive way. However, becoming fully engaged in such a program is not without cost. There are a number of ways to reduce both the cost and the risk associated with this kind of responsibility.

Perhaps the best way to minimize the risks involved in handling and storing hazardous materials and hazardous wastes is to keep as little of them on site as you possibly can. The less there is, the easier management and control will seem.

Reducing inventory can also diminish the chances of a catastrophic incident. It might even be possible for some employers to reduce the quantity of some of the more hazardous chemicals to below threshold levels. This reduction can be accomplished in a number of different ways, including ordering smaller shipments and maintaining the minimum inventory necessary for efficient and safe operation.

If you are not able to reduce inventory, you might consider storing quantities of certain of these substances in different locations around the shop. A failure, release, or spill in one area of the shop will not have an impact on the same chemical stored in another location and can prove a relatively practical and inexpensive way to reduce the risk of a catastrophic incident occurring.

Recognize, however, that all of this can seem easier said than done when you start to consider all the different kinds of materials and wastes we use on a daily basis, including liquid lubricants, glycol-based coolants, penetrating oils, chemical solvents, cleaners, and refrigerants, etc.

CONTAINMENT

We can't really ignore our responsibilities regarding hazardous substances released into the workplace or into the environment by spilling "from a container that will expose employees to the hazards of the materials ..." We are, in fact, responsible for implementing a program that has as its ultimate goal the containment and control of the spilled material.

The most effective and efficient controls and containment methods generally result in the creation of a barrier around the spill that works something like a dike, or—if the circumstance allows—by the creation of a ditch to contain the spill. The use of diatomaceous earth as an absorbent has proven equally effective over time. Recently, however, new products have come into the marketplace that can enhance the effectiveness of these traditional methods.

These containment agents can rapidly be uniformly applied over the contaminated area, or used to build a small dam. Applying the agents in this way can help to afford you the maximum level of effectiveness and response and almost immediate control of the spilled material. These agents can help solidify spilled liquids through absorption, suppress the vapors that can originate or emanate from those hazardous chemicals by covering the spill, or neutralize acids or caustics, mitigating the hazard.

The agents can be deployed by using a hand-held, portable pressurized applicator very much like a portable fire extinguisher, and/or by a nozzle and hose system very much like those used by emergency response teams that have become an integral part of most local fire departments. The advantage to these kinds of application devices is that they allow the operator to apply the chemical without having to come into direct contact with the spilled material.

Using absorption chemicals can help reduce the amount of time required for containment of the spill. If used correctly and directed at a run-off point or around the edges of the spill, the reactant solid will automatically create a barrier to slow or stop the spread of the material. Clean-up efforts are greatly reduced when solidifying agents, acid or caustic neutralizers, or activated carbon adsorbents are used. These agents can totally solidify, neutralize, or absorb liquid hazards when properly applied. The resulting substance is less hazardous. It is also easier to handle, transport, and dispose of.

The use of vapor suppression agents to help manage and control volatile emissions from hazardous substances, such as flammable liquids and those substances which present an inhalation hazard, is just as important for protecting workers. Applying this type of agent over the surface of the spill can knock down vapors almost instantly, providing the agent is appropriate for the hazardous material spilled.

It should be noted that some of the more effective agents can solidify the liquid flammable hazardous substances, while at the same time elevating the flashpoint above 140 degrees Fahrenheit, so the resulting substance may be handled as a non-hazardous waste under certain specific circumstances. It should also be noted that anyone doing this kind of clean-up work would be expected to wear the appropriate personal protective clothing and equipment and to follow the company's spill control policies and procedures. In addition, everyone must be educated, informed, and trained with regard to the nature of the chemicals spilled, the nature of the chemicals used in the clean-up, the care and use of the appropriate equipment, and the management and control of those hazards inherent with spill containment work.

EMERGENCY PLANNING

Planning is the process you must go through to achieve a predetermined outcome. One of its fundamental principles is the identification of as many of the necessary steps you will have to take in order to achieve the result you are looking for as possible. It also demands a realistic look at what can conceivably go wrong with your plan, so that these obstacles and pitfalls can be anticipated if necessary and avoided if possible. Everyone must know and understand what their specific responsibilities are with regard to this kind of planning, if all your pre-emptive efforts fail and the worst occurs despite all your efforts.

It takes at least a minimal amount of foresight to recognize what these potential dangers are, and a fair amount of discipline to prepare for them. It takes discipline, determination, and a little bit of skill to communicate this plan to everyone involved so that each person both knows and understands their place in it.

Because we are dealing with substances that are ignitable, corrosive, reactive, and toxic, any failure relating to the safe storage and handling of these materials can constitute an emergency.

e·mer·gen·cy (i-mur-jən-sē) *noun plural* e·mer·gen·cies **1.** A serious situation or occurrence that happens unexpectedly and demands immediate action. **2.** A condition of urgent need for action or assistance.

If such an emergency occurs, everyone in your organization should know what to do without having to wait for instructions or direction. The longer a hazard is allowed to go without the appropriate response, the greater the risk of exposure.

Emergency planning begins with an inventory of all the hazardous or potentially hazardous chemicals in the facility; repair procedures within the shop environment that could conceivably result in a hazard; the procedures involved in handling and storing hazardous materials or waste; how these hazardous materials or wastes are either used or generated; a risk analysis of potential failures; as well as an understanding of the many other dangers that could be associated with your environment at work. By anticipating what could go wrong, we can often mitigate that risk, if not eliminate it altogether.

Anticipating

If we are talking about chemical hazards, materials or waste, we can start with an evaluation of the overall risks inherent or associated with what we do. We can break those risks down into the kinds of chemicals used, the quantities of those chemicals stored onsite, and the nature of the containment or containers.

The chemicals we should be concerned with most are those substances that are poisonous or otherwise toxic, highly flammable, and/or unstable. In order to be effective, our analysis should always be focused on the regulatory requirements we are obligated to comply with, while our actions should be focused on the challenges inherent in responding to a worst case emergency situation.

That means the planning process should address at least the following areas:

- Planning for an emergency before it becomes an emergency
- Personnel with regard to their:
  - Awareness
  - Education
  - Training
  - Involvement
  - Responsibility
- Emergency preparedness and prevention
- Contingencies, especially with regard to:
  - Emergency exit routes or safe refuge within the facility
  - Decontamination

- First aid and/or other emergency medical procedures
- Communicating the emergency quickly and to everyone
- Emergency alarm and response procedures
- Personal protective equipment and other emergency response equipment
- Site map complete with hazardous materials and waste storage, containment and response gear, and evacuation route(s)
- Procedures for reporting an emergency to the appropriate government agency or agencies

While this list is certainly not everything you will need to consider while creating an emergency response plan, it is more than enough to get you started. However, there is another area of emergency planning to consider, and that is the planning required to prepare for an emergency that is human, and not substance-related.

Human Hazards

What do you do if there is an imminent threat of violence in the workplace? What do you do if there is a problem with some*one*, and not some*thing*?

What do you do if someone enters the workplace with a weapon, or threatens bodily harm to you or someone else? Will you know what to do, and will you react quickly enough to avert a disaster?

Your success is likely to depend more on common sense than it will on courage, and the odds are you won't have any of the right answers if you haven't considered the questions *before* you find yourself confronting a crisis.

Here are some important things to consider if the threat of workplace violence actually materializes:

1. Do not directly confront anyone who has a weapon.
2. Do not directly confront anyone who is acting in an irrational or aggressive manner.
3. Do not directly confront anyone who is acting in a belligerent, threatening, or confrontational manner.

If the level of violence remains verbal, or the threat remains non-physical, back away and disarm the situation as quickly as possible, if you can. Alert the remaining employees close to the situation, and then call the appropriate authorities. Retreating from a potentially dangerous situation may not seem like the bravest thing to do, but it just might be the most prudent. And by retreating you could conceivably avoid escalating the confrontation to the next, potentially more dangerous level.

SUMMARY

It is critically important for all of us to recognize the importance associated with handling and storing substances that are reactive, corrosive, toxic, or flammable (explosive or ignitable). Just defining these four characteristics should be enough to warrant meticulous attention to every detail associated with maintaining a safe and hazard-free environment.

However, for those of us who feel common sense isn't really common enough to adequately address these issues, a comprehensive set of policies and procedures will be required. These policies and procedures should address all the requirements relevant to your particular business.

Maximizing safety in the workplace, while minimizing the potential for workplace illnesses or injuries at the same time, can be accomplished by reducing both the number of hazardous materials stored in the facility as well as the quantity of each hazardous material.

Because of the danger involved in working with these hazardous materials, we must be prepared if our containment efforts fail and the threat of wholesale contamination materializes. That means recognizing that it *could* happen and being prepared if it *does* happen.

Finally, we must recognize that not all safety hazards in the workplace are the result of chemical exposure or unsafe practices. Some of the hazards we are likely to confront are human hazards; and while they may be less exotic, they can be every bit as dangerous—if not more so.

What do you think you can do to plan and prepare for the kinds of hazards you are likely to encounter in the workplace? Is there a difference between the danger involved in (a) handling hazardous materials, and (b) dealing with an angry and frustrated employee threatening violence or practicing abuse? What kinds of policies and procedures do you think it will take to create the kind of environment you would be proud of—an environment free from the threat of physical violence, verbal abuse, or chemical danger?

Write your answers in the chart provided on the following page.

Thoughts

Actions

Results

CHAPTER 7

Conclusion

I would be lying to you if I told you how excited I was in anticipation of putting this book together. Frankly, it just wasn't something I was looking forward to. Like almost every other shop owner I know, the regulatory requirements surrounding "Right to Know" and "Hazardous Communications" were just another example of the government involving itself where it seemingly has no real right to be. Aside from that, like almost all government rules and regulations, wading through the paperwork is almost as exciting as watching your favorite pet nap.

But something interesting—something fascinating, in fact—took place as I waded through the hundreds of pages of related materials I pulled off the web or acquired from other sources as I was researching safety communications. I began to understand the importance of these rules and regulations. Perhaps more important, I began to realize why compliance is critical, and that almost every automotive repair shop in the country is operating in violation of not only the letter of the law but the intent and spirit of the law as well.

As I researched the responsibilities we are all required to accept as shop owners and managers in this industry, I thought back to when there weren't any rules or regulations regarding safety in the workplace. I thought about the various injuries I suffered as a result of my own lack of awareness or respect for the dangers involved in servicing, maintaining, or repairing today's passenger cars and trucks, and I wondered how much wear and tear on my own body I could have spared myself. More than that, I thought about the emotional anxiety I could have avoided putting my wife through with the "Mitch will be a little late getting home this evening. The emergency room doctors think they'll have him all stitched up and ready to go home in another couple of hours and they think they got all the metal splinters out this time!" phone calls she received too many times in the course of our marriage.

For the first time, perhaps, I believe I really do understand what is expected of all of us and why it is so important to actively and aggressively communicate both the importance *of* and the need *for* a safe and hazard-free work environment.

Having said that, I must add that I have also gained a new appreciation not only for what that means, but also for what it requires of all of us, especially when we are talking about balancing rights against responsibilities.

I guess you can say that I have become a living example of the "Four Stages of Knowledge" we discussed at the very beginning of this book. At first, I really didn't know what I didn't know. I was ignorant—blissfully ignorant. I didn't know or understand what was required of me, although I had all the appropriate (generic and therefore non-applicable) materials on a shelf at the shop. And I didn't know or understand what I was actually responsible for with regard to communicating safe practices and risk avoidance to the people who worked both with me and for me.

After all, why should you have to tell anyone to be careful? Isn't being careful in everyone's best interest? If it is, why should you have to do anything? Common sense should take care of it all!

As I began to work my way through all the issues revolving around hazard and safety communications, I began to recognize that there were huge gaps in both my knowledge and my understanding.

I passed through and out of the door marked "ignorance" and into the room reserved for wisdom. I began to realize that the reason we talk about common sense as much as we do is because it isn't very common at all, and that as entrepreneurs, leaders, and managers in the industry we serve, we have a responsibility to communicate what we know and a responsibility to protect those around us from what they don't know.

The third stage of knowledge, not knowing or understanding what you do know, is a critical part of this discussion as well. The more information we provide to those around us, the more likely it is that they will be able to react quickly and correctly to an emergency without losing the precious seconds it takes to think your way through the situation. It's almost critical that you know what to do instinctively, almost subconsciously.

The fourth stage of knowledge, knowing what you know, is probably where all of us need to be, especially with regard to workplace safety. We have a responsibility to act quickly and decisively if we are forced to confront a disgruntled employee and there is the possibility that that confrontation could result in violence in the workplace. We need to know what to do if a waste oil container ruptures or your transmission flushing machine's holding tank ruptures.

There isn't a lot of time to think about what you need to do *after* there is an event or an accident, but there is an abundance of time to consider what the appropriate response might be *before* it happens or in the anticipation of it happening.

I suppose the worst part is—compliance isn't impossible. It isn't even all that difficult. But it does require effort and energy. Compliance does require commitment. What I have tried to do here, in the pages of this book, is to create for you both the knowledge of what is required of you, and why, as well as a broad overview of how to actually get the job done. The rest is up to you, just as it is up to me.

By the time you read this, I hope, I will have completed overhauling our own Safety Communications program here at the shop. Training will be on-going and comprehensive. Everyone in the shop will know where the hazardous materials and/or wastes are stored and how they should be handled, and everyone will know and understand what actually constitutes a hazard. Everyone will know and understand the What and the Where of using personal protective gear as well. They will know what gear is appropriate for the hazards associated with each of the tasks or operations we are likely to be called upon to confront, they will know where to find that gear, how to care for it, and when to replace it as well.

You see, I have come to recognize that where there are rights, there are responsibilities; and that responsibilities are just one of the costs associated with entrepreneurship and being your own boss.

So it really doesn't matter what we call you or even how we describe you—owner, manager, boss, capitalist, entrepreneur, or leader. What we are really talking about is a certain undeniable responsibility that comes with the territory, an intrinsic part of the job description for those willing to risk everything for the rewards of independence and interdependence.

MORE TO COME

This series, *Automotive Service Management*, is divided into three general areas of concentration and eight volumes. As we make our way through these eight volumes, we will be looking at virtually everything involved in successfully operating a contemporary automotive service facility in today's dynamic and challenging marketplace. That means we will be looking at everything from hiring practices, productivity, and service management to the cost of doing business (factored for service bay productivity), cost/value ratios, value-added service, and hazardous materials communication. As well as exploring our relationship with the consumer, our suppliers, our employees, and our industry, we will also be looking at promotions, advertising, sales, and marketing.

Our goal is to create the most comprehensive resource for automotive shop management available anywhere—a complete and exhaustive look at all the hard and soft skills required to manage any business, as well the tools required to run an efficient, effective, customer-responsive automotive service facility.

The series was not conceived or formulated in a vacuum. It did not originate in theory first, and then find its way into reality. It is the result of a lifetime spent in the automotive industry, backed up by four generations of genetic memory. I invite you to take whatever you can from these pages. Take the key performance indicators you find here and create a "snapshot" of your business as it is at the beginning of this journey. Then see how integrating what you find within the pages of these volumes changes those numbers, how it impacts the bottom line—your bottom line. And then let us know how you're doing!

In addition to this text, *Safety Communications*, there are seven other titles for you to read, enjoy, and work through: *Total Customer Relationship Management, From Intent to Implementation, Managing Dollars With Sense, Operational Excellence, The High Performance Shop, Building A Team*, and *Operations Management*.

If you have any suggestions for subsequent subjects, or ways we can improve what we have already presented, let me know. My goal is provide you with answers to the questions I had to search for or create myself. My goal is to ensure that you don't find yourself struggling the way I had to. My goal is to see that you are richly rewarded for your efforts and your ability. My goal is to see you succeed where others have failed.

So read this shop operations guide one section at a time, one volume at a time, or all at once. See what you like. See what you aren't all that comfortable with. *Then read it again!* Find those principles that make sense to you. Create a plan to integrate them into your business. *Then read it again!* Make the changes required to take your business to the next level. Monitor your progress. Document everything. *Then read it again!*

If it works for you, let us know. If you can find ways to help us help you make it work more effectively, let us know that too!

REFERENCES

U.S. Department of Labor, Occupational Safety and Health Administration website: *www.osha.gov.*

Personal Protective Equipment, US Department of Labor, Occupational Safety and Health Administration, OSHA 3077, 1998 (Revised).

April 1996, Todd Hamilton, NAEDA Safety Consultant/Federated Insurance, from NAEDA Equipment Dealer

www.omeda.org/fastfacts/5210.htm

www.osha.gov/SLTC/smallbusiness/index.html

www.osha.gov/SLTC/index.html#A

www.osha.gov/as/opa/worker/rights.html

www.osha.gov/pls/oshaweb/owadisp.show_document?p_table=STANDARDS&p_id=10654&p_text_version=FALSE

For more information about OSHA and workplace safety, please take a moment to contact J. J. Keller at 1-800-327-6868, or visit *www.jjkeller.com.*

Delmar Learning's OSHA-Net: Network/Internet-ready OSHA Training Programs

To purchase the following Delmar materials please visit *www.trainingbay.com*, or call 1-800-477-3692.

Personal Protective Equipment	ISBN: 07668-1812-8
Emergency Procedures	ISBN: 07668-1800-4
Hazard Communications	ISBN: 07668-1805-5
Respiratory Protection	ISBN: 07668-1814-4
Spill Prevention, Control and Countermeasures	ISBN: 07668-1816-0
Noise Control and Hearing Conservation	ISBN: 07668-1811-X
Self-Contained Breathing Apparatus	ISBN: 07668-1815-2
Environmental Awareness	ISBN: 07668-1802-0
Bloodborne Pathogens	ISBN: 07668-1798-9

SPECIAL THANKS

Charles R. Pearson, Jr, AAM, Special Staff Assistant

Howard W. Ferris, Director for Automotive Programs

Ashland Technology Center

MassBay Community College

Science and Advanced Technology Institute, 250 Eliot Street, Ashland, MA 01721